AF512205

CONCEPTS OF DATABASE SYSTEM

P.K. YADAV
Deptt. of Computer Science & Engineering
KIET, Ghaziabad

A.K. GOYAL
Deptt. of Computer Application
KIET, Ghaziabad

S.K. KATARIA & SONS®
Publisher of Engineering Books
4885/109, Prakash Mahal, Dr. Subhash Bhargav Lane,
Opposite Delhi Medical Association, Daryaganj, New Delhi–110002 (INDIA)
Phone: +91-11-23243489, 43551243; *Mobile:* +91-9871775858
e-mail: katariabook@gmail.com, katariabooks@yahoo.com
Website: www.skkatariaandsons.com
(e-books, subscriptions and bulk buying also available on our website)

Published by:

S.K. Kataria & Sons®

4885/109, Prakash Mahal, Dr. Subhash Bhargav Lane,
Opposite Delhi Medical Association, Daryaganj, New Delhi-110002 (INDIA)
Phone: +91-11-23243489, 43551243; *Mobile:* +91-9871775858
e-mail: katariabook@gmail.com, katariabooks@yahoo.com
Website: www.skkatariaandsons.com

First Edition: August, 2009-10; Third Edition: February, 2015, Reprint: 2024

Laser Typesetting by : Plus Computers, Delhi

Printed at : New A.S. Printing Press, Delhi

PREFACE TO THE THIRD EDITION

It gives me immense pleasure in presenting the third edition of the book on Database Management System. The DBMS has become very imoportant area in every commercial and engineering field.

This book dwells on fundamentals that one must learn in order to pursue virtually any endeavour in DBMS field. It trains student with the concepts needed to lay a Solid foundation for joining this exciting field. It is important to learn the fundamentals now so that one has the tools necessary to learn newer, emerging fields as they become more common. In simple words, reading this book will benefit anyone who wishes to arm themselves with substantive database skills.

The book has been presented with the view of databse applications in the various fields such as basic concepts of DBMS, Entity Relationship Diagram, Relational database, Normalization, SQL and PL/SQL, distributed database etc.

The present edition has been thoroughly revised and a lot of useful material has been added. To make it more understandable, Several solved examples and problems have been added from the point of view of examinations. The treatment of subject matter is kept quite simple and concepts are explained with the help of examples and illustrations.

The mistakes and misprints, which were recover in the previous edition, have been eliminated in this edition.

For any suggestions and comments about this book, please, kindly feel free to send your e-mail at **katariabooks@yahoo.com**.

Hope you will enjoy reading this book as much as I haved enjoyed writing it.

Pramod Kumar Yadav
Amit Kumar Goyal

PREFACE TO THE FIRST EDITION

It gives us immense pleasure in presenting the first edition of the book on "Concepts of Database System". This book dwells on fundamentals that one must learn in order to pursue virtually any endeavour in DBMS field. It trains student with the concepts needed to lay a solid foundation for joining this exciting field. The present book "Concepts of Database System" is meant for the 5th Semester Engineering students of B.Tech Computer Science, Information Technology and Electronics & Communication Branches, according to the new syllabus of GGSIPU, Delhi. This subject is also taught in 2nd Semester of M.C.A. Course & BCA Course of Guru Gobind Singh Indraprastha University, Delhi. It covers all the basic concepts of DBMS, including ER-diagram, RDBMS, Normalization, SQL & PL/SQL etc. This book is the outcome of continuous persuation of our friend and colleagues and from different engineering colleges. Institutions to develop basic course, material for this subject. We have tried to incorporate the suggestions of our colleagues and teachers.

We have tried our best to organise the contents of this book with sufficient solved examples and diagrams as well as previous years examination papers. To make it more understandable, several solved examples and problems have been added from the point of view of examination.

Hope you will enjoy reading this book as much as we haved enjoyed writing it.

Pramod Kumar Yadav
Amit Kumar Goyal

ACKNOWLEDGEMENT

I would like to acknowledge my deep sense of gratitude to Dr. Ajay Sharma, Director General, KIET, a man of Spiritual personality, for his continuous encouragement, effort towards achieving **UPTU Academic Excellence Award-2008**.

I would also like to thank Dr. Narendra Kumar, Director KIET, who has always motivated to achieve high academic values and has always been my source of Inspirations. I would like to thank Prof. **A.K. Sinha,** Dean (Academics) KIET, for acknowledging the work on the subject, his encouragement and critical suggestions.

I would like to express my special gratitude to **Dr. A.K. Sharma (HOD CSE Deptt.)**, Prof. Sanjeev Yadav, Prof. Adesh Pandey, Prof. Saurabh Chandra, Prof. A. P. Shukla, Prof. Vineet Sharma, Prof. Sumit Kumar, Prof. Arvind Kumar, Dr. Rabins Porrwal, Mr. Sachin Ahuja, Mr. Ankur Bhardwaj, Prof. Amit Goyal, Mr. Rajeev Chaurasiya, of KIET for their constructive suggestions.

I am also extremely indebted to Prof. N.C. Sharma, Dr. Yashpal Singh, **Dr. R.S. Yadav**, Prof. B.S. Roohani, Prof. Abhay Bansal, Prof. Vijay Shukla, Dr. Pankaj Agrawal for continuous encouragement and necessary help whenever required.

I am very thankful to my colleagues **Mr. Arun Yadav**, Mr. Lokesh Jain, Mr. Prabhat Kumar yadav, Mr. Dharmendra Tripathi, Mr. Nitin Goel, Mr. Ankit Vishnohi, Mr. Raj Kumar Goel, Mr. Ajeet Kr. Yadav, Mr. Chhote Lal Prasad Gupta, Mr. Sunil Vishwakarma, Mr. Piyush Bhushan Singh, Mr. Mohd. Islam, Mr. Narendra Kumar, Mr. Sunil Yadav, Mr. Arvind Dagur, Prof. Vikas Goel, Mr. Santosh Kr. Soni who have been the driving force behind this effort.

I would like to express my special gratitude to my parents Shri Shanker Prasad Yadav and Smt. Bella Devi for their perpetual love and concern. Their contribution to this book cannot be summed up in words.

I would like to express my heartiest thanks to my wife Suman Yadav, daughter Akansha Yadav, Son Yuvraj Yadav and younger brother Vinod Yadav for their patience and continuing support to make this task complete with pleasure.

I thank the staff of M/s S.K. Kataria & Sons especially to Mr. Sanjeev Kataria for his deep involvement in this project and bringing this text in presentable form. At last, I thank every one who has directly or indirectly contributed to this work.

Lastly, I am thankful to Mr. Manoj Tyagi and his staff from Plus Computer's for their hardwork.

Pramod Kumar Yadav
Amit Kumar Goyal

DEDICATED TO
OUR
PARENTS

CONTENTS

Chapter

1 Concepts of DBMS

Inside this Chapter

1.1. AN OVERVIEW OF DATABASE MANAGEMENT SYSTEM

The term database is made up of two separate words, *i.e.*, data and base. Database is a base for data, *i.e.*, an assembled group of data. A database allows easy and efficient storage, retrieval and modification of data, regardless of the amount of data being manipulated, esentially, database is a computerised record keeping system.

A database is a logically coherent, organised collection of similar data similar data refers to the collection of data, which is stored based on the same context. For example, an Employee database contains 'similar' data. For all Exployees and every Employee's entry contains a similar type of information. The organised information (*i.e.*, database) serves as a base from which the desired information can be retrieved, conclusions can be drawn and decisions can be made, by further reorganising or processing this data. In essence, a database not only stores information, but the information is integrated so that when the database is interrogated, users can get something useful from it.

"A Database Management System (DBMS) is a collection of interrelated data and a set of programs to access those data." The collection of data, usually referred to as the database, contains information relevant to an enterprise. The primary goal of a DBMS is to provide a way to store and retrieve database information that is both convenient and efficient.

Database is the combination of two words :

Database + Management System = DBMS

A database is a collection of related information stored, so that it is available to many users for different purposes.

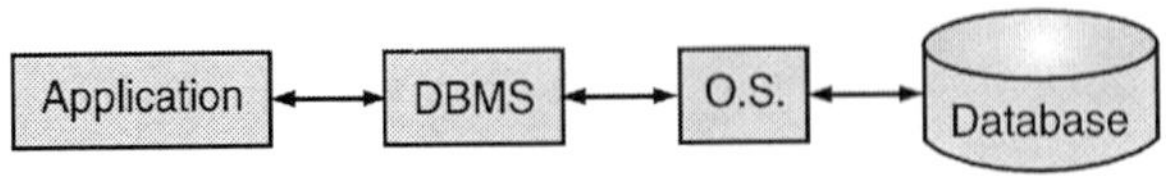

Fig. 1.1.

On the other hand management system is a collection of programs that enables users to create and maintain the database.

DBMS can also be define "as an interface between the application program and the operating system to access or manipulate the database".

A DBMS is a software system that allows access to data contained in the database. It's objective is to provide a convenient and effective method of defining, storing and retrieving the information contained in the database. The DBMS interfaces with application programs so that the data contained in the database can be used by multiple application and users.

The primary goal of a DBMS is to provide an environment that is congenial and efficient to retrieve and store information. It allows guest to store, update and retrieve data in abstract terms. DBMS eases the maintenance and retrieval of information from a database. A DBMS also relieves the user from having to know about the exact physical representations of data and from specifying detailed algorithms for sorting, updating and retrieving the data.

Principally a database system is a computerised record keeping system whose overall purpose is to maintain information and to make it available on demand. The management of data involves defining the structure for information storage and provision of mechanisms for its manipulation. DBMS also provides mechanisms for the safety of the stored information from data corruption and or unauthorized access. In addition, the database system avoids possible anomalous results in case the data is to be shared among several users.

We frequently use a database in our everyday life. Dictionary is one of the most common examples of a database where words are arranged alphabetically. Similarly, a telephone directory is another example of a database, where the names of telephone subscribers are listed in alphabetical order. Many people keep a list of the addresses of their friends and relatives which is usually written down in an address book. This is infact a database of the addresses. Based on this address database, several operations are performed, for example, if a person moves to a new location, his/her address is changed in the address book. Sometimes new addresses are also added in the address book.

1.2. NEED FOR DBMS

Historically, information systems employed stand-alone systems for separate application each with their own set of files. This implied the data such as addresses, may be duplicated in many separate systems, such as an employee's address may be held once in the payroll system and again in the personal system. This causes waste of space and inconsistency. There may be a possibility for the address to be updated in one system but not in the other, hence, making the data in consistent.

The problems described above led to development of the database approach. In the database approach, an organisation attempts to build an integrated corporate database to support all the data requirements of the organisation. DBMS are very good at organising and maintaining large collections of persistent data. A DBMS is like a suitcase where all the stuff (data) is put so that it is in one place and easy to reach. This protect data from unauthorised access and accidental corruption or loss due to hardware and software failures. DBMS allows concurrent access, meaning that a single data set can be accessed by more than one user at a time.

1.3. COMPONENT OF DBMS

Usually a DBMS is a large software package that carries out diverse tasks including the provision of facilities to enable the user to access and modify information in the database. On one hand it is intermediate link between the physical database, the computer and the operating system and on the other hand, the users.

> **DO YOU KNOW?**
>
> A database manager is a program, which provides the interface between the physical database and the application programs and queries submitted to the system.

Essentially, this system involves data, the hardware that physically stores that data, the software that utilises the hardware's file system in order to store the data and provide a standardised method for retrieving or changing the data and finally the users who turn the data into information.

1. **Uses.** In DBMS, generally three broad classes of users are considered. These are application programmers, end users and DBA (Database Administrator). The application programmers develop the application programs. There programs can manipulate the database in all the possible ways. The end users access the database from a terminal using a query language provided by the database system or through application programs developed by application programmers. The database Administrator is the person who is responsible for the design, construction, and maintenance of a database.
2. **Software.** Software of a database management system includes the DBMS, operating system, network software (if necessary) and the application program.

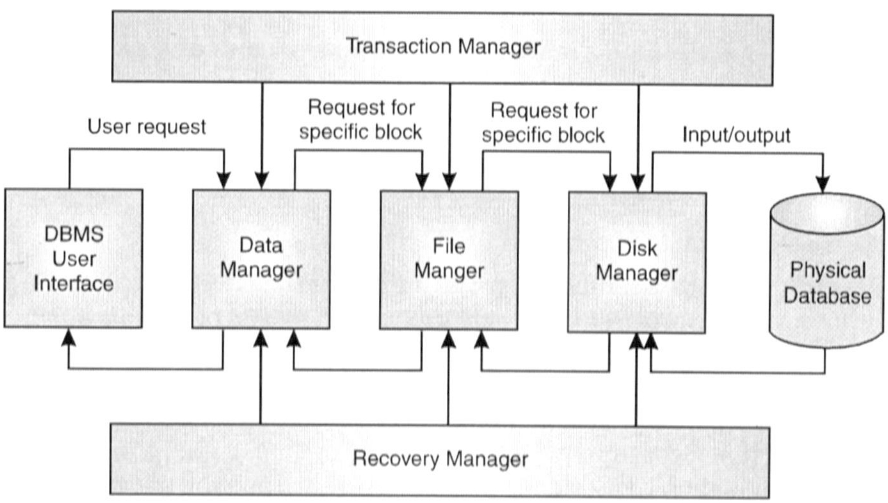

Fig. 1.2. DBMS components

3. **Hardware.** Hardware of the system can range from a PC to a network of computers. It also includes various storage devices (like hard disks) and input and output devices (like monitor, printer etc.)
4. **Data.** Data stored in a database includes numerical data including whole numbers and floating-point-number and non-numerical data such as characters, data or logical (true or false). More advised systems may include none complicated data entities such as printers, images and data types.

Above diagram shows a typical interaction among various DBMS components. The user requests for specific information with the help of user interface, that is application programs. This request is processed by DBMS data manager and after processing, data manager requests for the specific records to the file manager. The file manager then requests for the specific block to the disk manager. The disk manager then retrieves the block and sends it to file manager, which sends the required record to data manager. The transaction manager supervises the data transaction that is carried out between the data Manager, and the disk manager. The recovery manager keeps a check on the transacted data so that in case of system failure, the data can be protected.

1.4. APPLICATIONS OF DBMS

Some of the important application fields of DBMS are :

1. **Banking :** For maintaining customer information, accounts, loans and banking transactions.
2. **Universities :** For maintaining student records, course registration and grades.
3. **Railway Reservation :** For checking the availability of reservation in different trains, tickets, etc.
4. **Airlines :** For reservation and schedule information.
5. **Telecommunication :** For keeping records of calls made, generating monthly bills etc.
6. **Finance :** For storing information about holidays, sales and purchase of financial instruments.
7. **Sales :** For customer, product and purchase information.

1.5. ADVANTAGE OF DBMS

Some of the major advantages provided by the DBMS are :

1. **Reduction in Data Redundancy :** Data redundancy refers to the duplication of data. In non-database systems, each application has its own separate flies. This can often lead to redundancy in stored data, which results in wastage of space. For example, in case of file-based system of a magazine publishing house, the customer department keeps a separate customer details file. A similar file is also kept with the postage-handling department. Hence the same data may not be updated consistence. Suppose a customer, requests for change in his/her posting address to the customer department and the changed address is reflected in his/her record. However, it is not guaranteed that the same change is also done in the postage–handling department's life. If the database system forgets or delays to update this file, this may lead to postage of the subscribed magazine to the old address.

A DBMS does not maintain separate copies of the same data. All the data is kept at one place and various applications refer to the data from this centrally controlled system. As a result, when the customer demands the change in address, it is done at one place from where the concerned department's application uses the data to process.

2. **Reduction in Inconsistency:** This is a corollary of the previous point discussed above. It is obvious that if the given fact is represented by a single entry then such inconsistence may not occur. Alternatively, even if the redundancy is not removed but is controlled, then to the DBMS ensures that the database is always consistent by ensuring that any change made to either of the two entries is automatically applied to the other one also. This process is also known as propagating update.
3. **Sharing of Data:** Sharing of data allows the existing applications to use the data in the database. It also helps in developing new applications, which will use the same store data. Due to shared data only, it is possible to satisfy the data requirements of new applications without having to create any additional stored data or with marginal modification.
4. **Enforcement of Standards.** Since access to the database must be through the DBMS, Standards are easier to enforce. Standards may relate to the naming of the data, the format of company or organisation using the database. Standardization of data particularly helps in migrating or interchanging data between different systems.
5. **Improvement in Data Security.** In conventional systems, applications are developed in an improper manner, usually different systems of an organisation would access different components of the operational data. In such an environment, enforcing security can be quite difficult. Setting up of a database management system makes it easier to enforce security restrictions since the data is stored centrally. DBMS can ensure that the only means of access to the database is through the authorised channel. Hence, data security checks can be carried out whenever access is attempted to sensitive data. To ensure security, DBMS provides security tools such as user name and passwords.
6. **Maintenance of Data Integrity.** Data integrity refers to ensuring that the data in the database is accurate. Since, in DBMS, the data is centralised and is used by a number of users at a time, it is essential to enforce integrity controls. Integrity may be compromised in many ways.

   However, since all data is stored only at one place, it is often easier to maintain integrity then in conventional system.
7. **Better Interaction with Users.** As compared to traditional database system, a DBMS often provides better service to the users. In conventional systems, usually the information is poorly arranged and as a result, the availability of up-to-date information becomes poor. However, in case of DBMS, the availability of up-to-date information improves since the data can now be shared and the DBMS makes it easy to respond to unforeseen information requests.
8. **Efficient System.** It is very common to change the contents of stored data. These changes can easily be made in a database management system, than in a conventional system as these changes do not need to have any impact on application programs. The cost of developing and maintaining systems is also lower.

1.6. DISADVANTAGES OF DBMS

1. **Problems Associated with Centralization :** Centralization increases the security problems and disruption due to the downtimes and failures.
2. **Cost of Software :** Today's there are several software which are very costly. Hence from economic point of view it is the drawback.
3. **Cost of Hardware :** It incurred in the application of DBMS is its major disadvantage. Cost of the H/W is also one of the major drawback.
4. **Complexity of Backup and Recovery :** DBMS provides the centralization of the data, which requires the adequate backups of the data so that in case of failure, the data can be recovered. Hence backup problem is also the drawback.

1.7. CHARACTERISTICS OF DATABASE APPROACH

In the database approach, a single repository of data is maintained that is defined once and then accessed by various users. In contrast, in a databse, the names or labels of data are defined once and used repeatedly by queries, transactions and applications.

The main characteristics of databse approach are as follows :

1. Self-Describing Nature of a Database System.
2. Insulation between programs and Data and Data Abstraction.
3. Support of multiple views of the data.
4. Sharing of data and multiuser transaction processing.

1.7.1. Self-Describing Nature of a Database System

The main characteristics of the database approach is that the database system contains not only the database itself but also a complete definition or description of the database structure and constraints. This definition is stored in the DBMS catalog called meta-data and it describes the structure of the Primary database.

The catalog is used by the DBMS software and also by database users who need information about the database structure. A general-purpose DBMS software package is not written for a specific database application. Therefore, it must refer to the catalog to know the structure of the files in a specific database such as the type and format of data it will access.

1.7.2. Insulation between Programme and Data and Data Abstraction

The characteristics that allows program data Independence and program-operation independence is called data Abstraction. A DBMS provides users with a conceptual representation of data that does not include many of the details of how the data is stored or how the operations are implemented. Informally, a data model is a type of data abstraction that is used to provide this conceptual representation.

In database approach, the detailed structure and organization of each file are stored in the catalog. In object-oriented and object-relation databases, the abstraction process, includes not only the data structure but also the operations on the data.

1.7.3. Support of Multiple Views of the Data

A database has many users, each of whom may require a different perspective or view of the database. A view may be subset of the database or it may contain virtual data that is derived from the database files but is not explicitly stored. A multi users DBMS whose users have a variety of distinct applications must provide facilities for defining multiple views.

1.7.4. Sharing of Data and Multiuser Transaction Processing

A multiuser DBMS, as its name implies, must allow multiple users to access the database at the same time. This is essential if data for multiple applications is to be integrated and maintained in a single database. The DBMS must include concurrency control software to ensure that several users trying to update the same data do so in a controlled manner so that the result of the updates is correct. A fundamental role of multiuser DBMS software is to ensure that concurrent transaction operate correctly and efficiently.

The concept of a transaction has become central to many database applications.

1.8. DATABASE SYSTEM CONCEPTS AND ARCHITECTURE

DATA ABSTRACTION

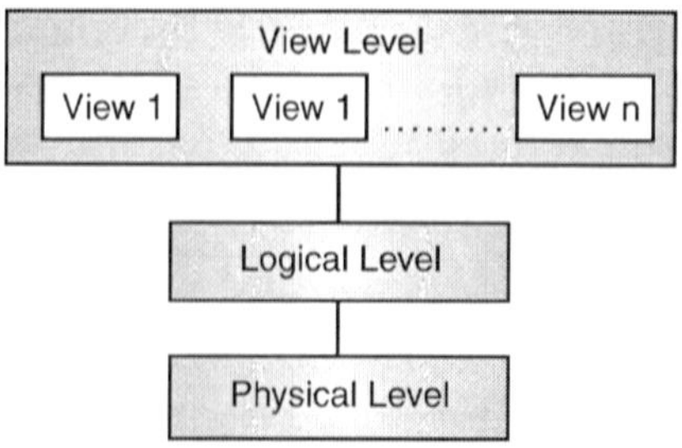

Fig. 1.3. The three levels of data abstraction.

1. **Physical Level :** This is the lowest level of abstraction and describes how the data are actually stored. The physical level describes complex low-level data structures in details.
2. **Logical Level :** This is the next-higher level of abstraction and it describes what data are stored in the database, and what relationship exist among those data. The logical level thus describes the entire database in terms of a small number of relatively simple structures.
3. **View Level :** This is the highest level of abstraction describes only part of the entire database. Even though the logical level uses simpler structures, complexity remains because of the variety of information stored in a large database. The view level of abstraction exists to simplify their interaction with the system. The system may provide many views for the same database.

1.9. DATABASE MODELS

Data Modelling is a way of organising a collection of information pertaining to a system under investigation. A database model is a collection of conceptual tools for describing data, data relationships, data semantics and consistency constraints. It consists of two parts :

- A mathematical notation for describing the data and relationships.
- A set of operations used to manipulate that data.

Every database and database management system is based on a particular database model. A database model consists of rules and standards that define how data is organised in a database. It provides a strong theoretical foundation of the database structure. This theoretical base enhances the power of analysis, the ability to extract inferences and to create deductions that emerge from the new data. Different models provide deficient conceptualizations of the database and they have different outlooks and different perspective. There are four basic types of database models : Hierarchical, Network, Relational and object-oriented database model.

1.9.1. Hierarchical Database Model

The hierarchical database is the oldest form of database. It was developed by the IBM for its IMS (Information Management System) database. This data model organises the data in a tree structure, that is, each child node can have only one parent node and at the top of the structure, there is a single parentheses node. In this model, a database record is a tree that consists of one or more groupings of fields called segments, which makeup the individual nodes of the tree. All the instances of a specific record are grouped together as a record, the hierarchical model uses parent-child, that is, one-to-many relationship.

The main advantage of hierarchical database is that data access is quite predictable in structure, and therefore, both retrieval and updates can be highly optimised by a DBMS. However, as the parent-child relationship is one-to-may, it restricts a child segment to having only one parent-segment. In the hierarchical database model, the links are 'hard coded' into the data structure, that is, the link is permanently established and cannot be modified. The hard coding makes the hierarchial model rigid.

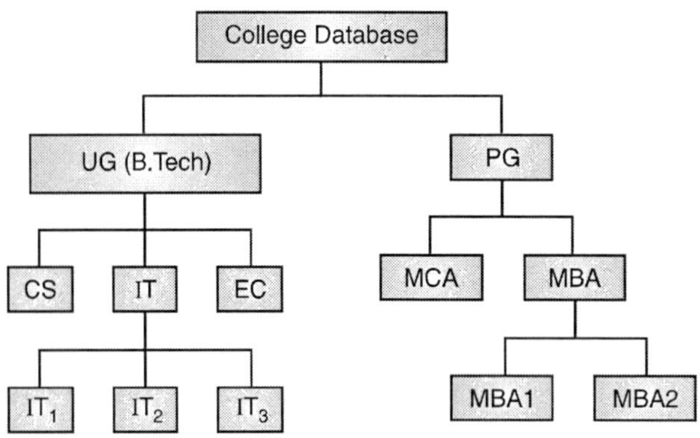

Fig. 1.4.

1.9.2. Network Database Model

The network database model was developed as an alternative to the hierarchial database. This model was formalised in 1971 by the Database Task Group (DBTG) of the conference on Data Systems Languages. The network database model expands on the hierarchial model by providing multiple paths among segments, that is, more than one parent-child relationship. Hence, this model allows having 1 : 1 (one-to-one), 1 : M (one-to-Many) and M : M (Many-to-Many) relationships.

The basic data-modelling construct in the network model is the set construct. A set consists of an owner record type, a set name and a member record type. A member record type can have that role in more than one set and as a result, multi-parent concept is supported. A networked database stores, information in data sets, which are similar to files and tables.

Although supporting multiple paths in the data structure eliminates some of the drawbacks of the hierarchical model, the network model is not very practical. The primary drawback of networked databases is that it can be quite complicating to maintain all the links.

In networks database model, all the relationships are hardwired (Pre-computed) and build into the structure of the database itself. Therefore, they are very efficient in space utilization and query execution times. However, the price for such performance is inflexibility and great difficulty of use.

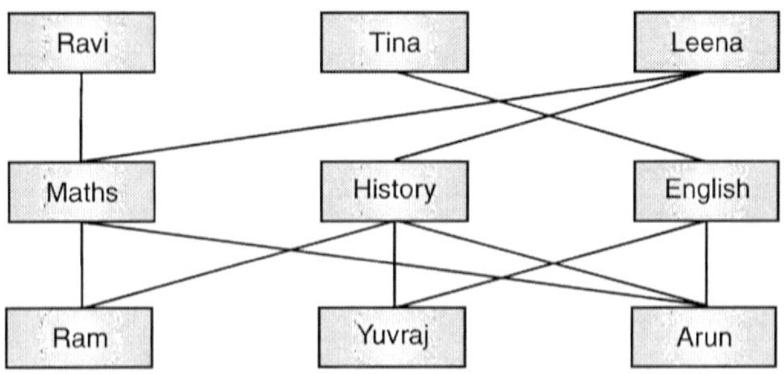

Fig. 1.5. Network database model.

1.9.3. Relational Database Model

The hierarchical and network database model, although more flexible than tradition file systems, were still not flexible enough. The limitations of these database systems led to the development of relational database management system (RDBMS). The key difference between previous database models and relational database model is in terms of flexibility. It is relatively easy and quick to create a new database structure and amend existing structures in relational systems as compared to reconstructing a network database which is a complex and time consuming task.

Relational database management system is based on the relational model developed by E.F. Codd. A relational database represents all datas in the database as simple two dimensional tables called relations that are the logical equivalent of files.

Each row (record) of a relational table, called tuple, represents a data Entity with columns of the table representing attributes (fields). The allowable values for these attributes are called the domain, indicating that values can be placed in each of the columns of the relational table. Each row in a relational table must have a unique primary keys and typically, it has some secondary keys, these secondary keys correspond with primary keys in other tables, *i.e.*, relate all the tables. For example, in Fig. 1.9 mentioned below, along with ISBN as the primary key, the Book table has secondary keys AUTHOR_ID and PUB_ID. In turn, these keys serve as primary keys for the AUTHOR and PUBLISHER tables, respectively.

AUTHOR TABLE

Author_ID	Author_Name	Author_DOB
567896	Jack	8/07/1956
123875	Kabir	9/04/1958
876923	Gaurav	10/08/1960
776544	Hars	9/10/1962

PUBLISHER TABLE

Pub_ID	Publisher	Pub_Address
8876432	89 bex Books	Walker street
994683	5 KG Pub.	Joint street
888777	Pearson Ed.	Mumbai
976543	Katson Pub.	New Delhi

BOOK TABLE

ISBN	Author_ID	Pub_ID	Date	Title
1-6385-81-11	567896	8876432	1995	Computer Fundamental
3-8674-10-21	123875	994683	1998	Operating System
8-6432-17-18	876923	888777	1978	M.S. Office
6-7431-22-15	776544	976543	1999	Data Structure

Fig. 1.6. Relational Database Model.

In a relational database, three basic operations are used to develop useful sets of data : selection, Projection and join. The selection operation retrieves certain records from a table based on the user specified criteria. The projection operation extracts fields from a table. Permitting the user to create new tables that contain only the required information. The join operation combines the data from the two tables based on a common column, providing the user with more information than is available in individual tables. Together these three operations are called relational algebra. The relational databases provides flexibility that allows charges to the database structure to be easily accommodated. The relational database system is truly a mathematically complete data model. Relational model facilities multiple views, that is, it is easy to present different user with different views of the same database. It also **in corporates concurrency** control, which guarantees the correct execution of concurrent queries.

1.9.4. Object-Oriented Database Model

The relational database model has been successfully in a wide variety of application areas. However it does not easily support the distribution of one database across a number of servers, which is the natural node for the internet. Due to these reasons, object-oriented database management systems (ODBMS) was developed. In these databases, the users can define own data access methods, the representation of data and the method of manipulating it. The objected-oriented database model is a relatively new database model and provides an outlook for the future database models. An object-oriented databases stores and maintains objects. An object is an item that can contain both data and the procedure that manipulation the data.

As compared to relational databases, object-oriented database system can store more types of data and access these data much faster. With an object-oriented database, a system can store and access unstructured data such as pictures, videoclips, sounds more efficiently.

Object-oriented databases often use a query language called Object Query Language (OQL) to manipulate and retrieve data.

In figure 1.10 sketch below is a simple example of object-oriented database structure. The class **population** is the root of a class hierarchy, which includes the **Nation** class. The **Population class** is also the root of two subclass, **Men** and **Women**. The **Nation** class is the root of other subclasses namely, **Country 1**, **Country 2** and **Country 3**. Notice that each class has its own set of attributes apart from the root class's attributes.

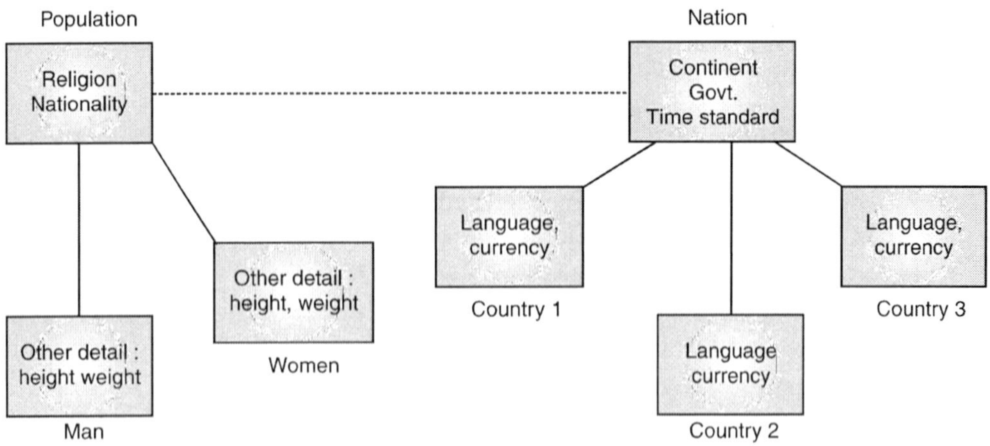

Fig. 1.7. Object-Oriented database model.

1.10. SCHEMA AND INSTANCES

"The collection of information stored in the database at a particular moments is called *an instance of the database*."

"The overall design of the database is called the *database schema*."

The concept of database schema and instances can be understood by analogy to a program written in a programming language. A database schema corresponds to the variable declarations in a program. Each variable has a particular value at a given instant. The values of the variables in a program at a point in time corresponds to an instance of a database schema.

There are three types of schema.

1. Physical schema
2. Conceptual schema
3. External schema

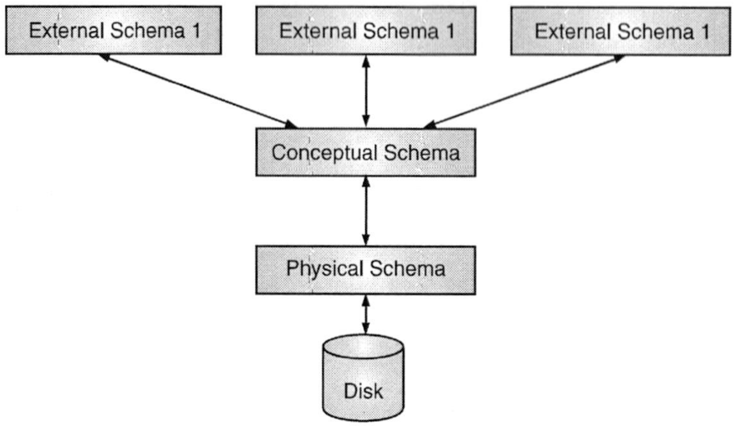

Fig. 1.8.

1. **Physical Schema :** Physical schema describes the database design at the physical level. It specifies additional storage details. Essentially, the physical schema summarizes how the relations described in the conceptual schema are actually stored on secondary devices such as disks and tapes.
2. **Conceptual Schema :** The conceptual schema (sometimes called *logical schema*) describes the stored data in terms of the data model of the DBMS. In a relational DBMS, the conceptual schema describes all relations that are stored in the database.
3. **External Schema :** External schemas, which usually are also in terms of the data model of the DBMS allow data access to be customized at the level of individual users or groups of users.

1.11. APPLICATION ARCHITECTURE OF DBMS

Basically there are two types of architecture of DBMS :

1. Two-tier architecture
2. Three-tier architecture

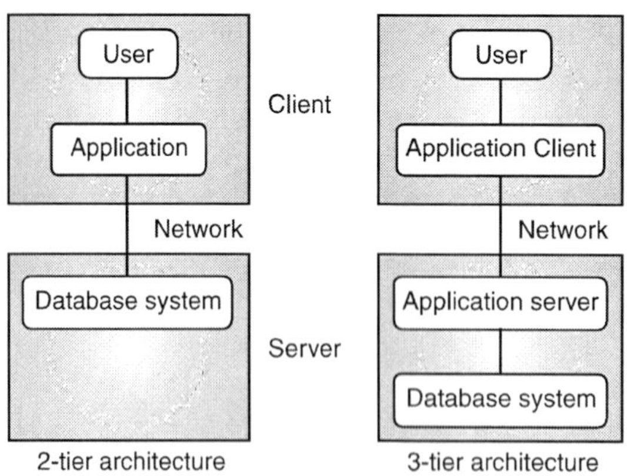

Fig. 1.9. 2-tier architecture and 3-tier architecture.

1.11.1. Two-Tier Architecture

The application is partitioned into a component that resides at the client machine, which invokes database system functionality at the server machine through query language statements. Application program interface standards like ODBC (Open Database Connectivity) and JDBC (Java Database Connectivity) are used for interaction between the client and the server.

1.11.2. Three-Tier Architecture

The client machine acts as merely a front end and does not contain any direct database calls. Instead, the client end communicates with an application server, usually through a forms interface. The application server in turn communicates with a database system to access data. Three-tier applications are more appropriate for large applications, and for applications that run on the world wide web.

1.12. DATA INDEPENDENCE

The ability to modify a schema definition in one level without affecting a schema definition in the next higher level is called *Data Independence.*

Data independence is one of the main advantage of the DBMS.

It is of two types :

1. Physical Data Independence
2. Logical Data Independence

1.12.1. Physical Data Independence

The separation of the conceptual view from the internal view enables us to provide a logical description of the database without the need to satisfy physical structures. This is often called physical data independence. Modification at the physical level is occasionally necessary in order to improve performance. Simply, it refers to the immunity of an application to changes in the internal mode and access strategy. Examples of physical independence are reorganisation of files, adding a new access path etc.

DO YOU KNOW?

Logical data independence is difficult to achieve than physical data independence because the application programs are very much dependent on the logical structure of the data.

1.12.2. Logical Data Independence

The separation of the external views from the conceptual views which enables the users to change the conceptual view without affecting the external views or application program is called logical data independence. Simply, it refers to the immunity of the external model to changes in the conceptual model. Examples of logical data independence are addition/removal of entities.

1.13. DATABASE LANGUAGE AND INTERFACES

In many DBMS, where no strict separation of levels is maintained, one language called the *data definition language* (DDL), is used by the DBA and by database designers to define both schemas.

In DBMS where a clear separation is maintained between the conceptual and internal levels, the DDL is used to specify the conceptual schema only.

Another language, the storage definition language (SDL) is used to specify the internal schema.

The third language the view definition languages (VDL) is used to specify user views and their mapping to the conceptual schema.

Data manipulation language (DML) is used for update, delete, insert, retrieve.

It is of two types :

1. High-level or Non-procedural DML.
2. Low-level or Procedural DML.

Non-procedural DML can be used on its own to specify complex database operations in a concise manner whereas procedural DML retrieves individual records or objects from the database and processes each separately.

DBMS INTERFACES

There are six types of interface provided by a DBMS :

1. Menu-Based Interfaces for Browsing
2. Forms-Based Interfaces
3. Graphical-User Interfaces
4. Natural Language Interfaces
5. Interface for Parametric Users
6. Interfaces for the DBA

Menu-Based Interfaces for Browsing : These interfaces present the user with lists of options, called *menus*, that lead the user through the formulation of a request. Pull-down menus are becoming a very popular technique in window-based user interfaces. They are often used in browsing interfaces, which allow a user to look through the contents of a database in an exploratory and unstructured manner.

Form-Based Interfaces : A form-based interfaces displays a form to each user. Users can fill out all of the form entries to insert new data, or they fill out only certain entries, in which case the DBMS will retrieve matching data for the remaining entries. Many DBMSs have forms specification languages, special languages that help programmers specify such forms.

Graphical-User Interfaces : A graphical user interfaces (GUI) typically displays a schema to the user in diagrammatic form. The user can then specify a query by manipulating the diagram. In many cases, GUI utilize both menus and forms. Most GUI use a pointing devices such as a mouse, to pick certain parts of the displayed schema diagram.

Natural Language Interface : These interfaces accept requests written in English or some other language and attempt to "understand" them. A natural language interface usually has its own "schema" which is similar to the database conceptual schema.

Interfaces for Parametric Users : Parametric users, such as bank tellers, often have a small set of operation that they must perform repeatedly system analyst and programmers design and implement a special interface for a known class of naive users.

Interfaces for the DBA : Most database system contain privileged commands that can be used only by the DBA's staff. These includes commands for creating accounts, setting system parameters etc.

DATA DEFINITION LANGUAGE (DDL)

It is a set of SQL commands used to create, modify and delete database structures but not data. These commands are normally not used by a general user, who should be accessing the database via an application. They are normally used by the DBA to a limited extent, a database designer or application are immediate developer. These statements are immediate, *i.e.*, they are not susceptible to Rollback commands. DDL also updates a special set of tables called the data dictionary or data directory.

A data dictionary contains metadata, *i.e.*, data about data. The schema of a table is an example of metadata. A database system consults the data dictionary before reading or modifying actual data.

Fundamental Commands of DDL :

The main tasks of the Data Definition Language are :

- To create the database objects like tables, Indexes etc.
- To modify the databse objects, for example, we can change the data type of a field or we can drop or add a field from/to the table.
- To destroy the database objects

Creating Database Objects :

For creating database objects we use the CREATE statement. For example :

Syntax : SQL > CREATE TABLE < table Name > (fieldname$_1$, Datatype$_1$, (width$_1$), fieldname$_2$ Datatype$_2$ (width$_2$), ...);

Example : SQL > CREATE TABLE STUDENT (NAME VARCHAR$_2$ (12), AGE NUMBER (2, 0), ADDRESS VARCHAR$_2$ (10), Class Number (2, 0));

Output : Table Created.

This Creates a table of students having field name, age, address and class.

Description of table

It is possible to obtain description of table to know field and field width by "describe" command.

Syntax : SQL > DESCRIBE Tablename; OR DESC Tablename;

Example : SQL > DESCRIBE STUDENT;

Output :

	TYPE
NAME	VARCHAR2(12)
AGE	NUMBER(2)
ADDRESS	VARCHAR2(10)
CLASS	NUMBER(2)

This describes the database of student.

Creating table using Constraint

Example : SQL > CREATE TABLE Dept.

(deptno.NUMBER(2) PRIMARYKEY, DNAME VARCHAR2(10) NOTNULL, LOC VARCHAR2(25) CHECK (LOC IN ('Delhi', 'Bombay', 'Noida')));

In the above example, Dept. table has three fields namely dept.no, dname and loc. The column deptno. will accept the number type of data with max width of 2. it has a Constraint Primary Key, means that we have to enter the unique deptno. for every valid department, which is to be stored in the database. The dname columnis not nullmeans that every new department will be having a valid name. We cannot leave this column as NULL. The last column, loc is of string type with a check constraints which allows only three locations namely 'Delhi', 'Bombay' and 'Noida'. Any value of other than these three will not be accepted.

Declarative Data Integrity

There are two types of data Integrities

1. Entity
   (*a*) DEFAULT – Specifies default value.
   (*b*) NOTNULL – Specifies that value should not be null.
   (*c*) UNIQUE – Specifies that value should be unique.
   (*d*) CHECK – Specifies that values should be ckecked.
2. Referential
   (*a*) PRIMARY KEY – The field is Primary field.
   (*b*) FOREIGN KEY –
   (*c*) DELETE/UPDATE Restrict –

NULL Value Concept

If a particular row do not contain a data, then it may be treated as null value.

Column of any data type may contain NULL unless, the column was defined as NOT NULL, When table was created.

NOT NULL

If a column is defined as NOT NULL, then it becomes a mandatory column *i.e.*, the user has to enter the data into it. It can take duplicate values if it is not a unique column but cannot take NULLS.

Example : SQL > CREATE TABLE Emp (EMP_NO VARCHAR2(2) NOT NULL);

Primary Key

A primary key is a one or more column used to uniquely identify each row in the table. These values should never be changed and should never be NULL. Through this key, rows in the table are individually identified. NULLS are not allowed in the primary key column. The important point about primary key is that only one primary key per table is allowed. The reason may be that the important task which the primary key performs is to uniquely identify the row.

Example : SQL > CREATE TABLE Emp (emp_no VARCHAR2(2) Primary Key);

In this case Emp_no is primary key.

Unique Key

This is similar to primary key. It is used to ensures that information in the column for each record in unique.

Example : SQL > CREATE TABLE Emp (emp_no VARCHAR2(2) Constraint (Unique));

In this case emp_no is unique key.

Default Value

When the cell is created, default value can be assigned to it as below.

Example : SQL > CREATE TABLE Emp (emp_no VARCHAR2(2) Default 0);

In this case default value of emp_no is 0.

Foreign Key

Foreign key represents the relationships between tables. This ensures the Referential Integrity Rule. A forgein key is a column or a group of columns whose values are derived from the primary key of the same or some other table. It can also be used elsewhere. A foreign key must have a corresponding primary key value in the primary key table to have a meaning. A foereign key value can be left null. It is inserted by.

REFERENCES emp 1;

In this case the Reference key in emp table is Foreign key.

Check Constraints

These constraints can be inserted to check integrity of the name data entered.

Syntax : CONSTRAINT CHECK (Logical Expression)

Example : SQL > CONSTRAINT CHECK (NAME LIKE "M%");

The constraint can be declared at the time of creations of table.

Example of Constraints :

```
SQL > CREATE TABLE LOANS (
account NUMBER (6, 0) NOT NULL,
Loan_number NUMBER (6, 0) NOT NULL UNIQUE,
Loan_type VARCHAR2(8)
   CONSTRAINT Loan_type
      in ('PERS', 'HOME', 'AUTO')),
amount NUMBER (8, 0) NOT NULL,
Loan_Date DATE DEFAULT SYSDATE);
```

Creating Index

Indexes are used to speed up the retrieval of records in response to certain search condition. The index structure provides the alternative way of accessing the records without affecting the physical, placement of records on physical media. The index is constructed based on a column or a group of columns from the table. When this index is used in the query for searching the records, it retrives the information in an efficient manner.

The Syntax

SQL > CREATE [UNIQUE] IMDEX <index name> ON <table name> ([column_name [ascending or descending] [column name [order] .....]);

Example : SQL > CREATE INDEX Empno_index ON emp (emp no);

Modifying the table

To change the definition of a table, we have to use the ALTER TABLE command. It has three clauses ADD, MODIFY and DROP, which are used to :

- Add a new column
- Add a new constraint
- Modify a column
- Modify a constraint
- Drop a column
- Drop a constraint

Syntax : SQL > ALTER TABLE <table name> {ADD/MODIFY/DROP} (column specifications);

Examples : SQL > ALTER TABLE emp ADD (Commission NUMBER (7, 2));

SQL > ALTER TABLE EMP MODIFY (Commission NUMBER (8, 2));

SQL > ALTER TABLE emp DROP COLUMN Commission;

Droping an object (table/view)

The DROP Command is used to drop an object permanently.

Syntax : SQL > DROP TABLE <object name>;'

Example : SQL > DROP TABLE emp;

DATA MANIPULATION LANGUAGE (DML) :

It is the area of SQL that allows changing data within the database.

A DML is a language that enables users to access or manipulate data or organized by the appropriate data model.

Through DML, we can

1. **Insert the new data in the database :** For example, if a person joins the company then we have to insert the information of that employee in our employee database.
2. **Modify the data, already stored in the database :** For example, in employee database, if an employee gets like (increment) in his salary than we have to modify the record of that employee.
3. **Deletion of the Data :** For example, if an employee region from the company than we have to delete that record from the database.
4. **Retrieve the information :** For example, if we want to display the address of employee than we have to write the query statement for that information.

It is of two types :

1. Procedural DML.
2. Non-procedural DML.

1. Procedural DML : It requires a user to specify what data are needed and how to get those data. It also requires writing the procedures/methods to specify what data is needed. Procedural DML language has all the features of any other high-level language including control constructs, error handling and other features.

2. Non-Procedural DML : It is also called as declarative DML and it require a user to specify what data are needed without specifying how to get those data. Non-procedural DML is the collection of commands for data manipulation including insert, update, delete and query the data.

Fundamental Commands of DML :

Inserting new records in the table :

The values could be inserted into the table by use of **INSERT INTO COMMAND**.

Syntax : INSERT INTO emp (emp no, name, department, age, bp) VALUES ("01, 'Suresh', 'Production', 35000);

Values must be listed in the same order in which the corresponding columns are specified. If any column is not specified in the list then it will take NULL corresponding to that column.

Example : SQL > INSERT INTO STUD (NAME, AGE, ADDRESS, CLASS) VALUES ('VIJAY', 24, 'KANPUR', 1);

Output : 1 row created.

This insert values into database of student.

Addition of field from the SQL editor :

It is possible for usa to allow SQL to read data values from the screen by attching "&" to the fields as below.

Syntax : Insert into table values (&field$_1$' &field$_2$' &field$_3$ .....);

The character fields are enclosed in single quote.

Example : SQL > Insert into emp values (&emp no, '&empname', '&address', '&desig');

When the above command is executed, the editor ask the value for given field, then the values are supplied from the data screen.

When we want to insert the records of 15 students or more than one students, than this method is more suitable. Once the value is inserted then just press 'back slash' ('/') and press enter, it will ask for another value. Similarly we can insert all the records of 15 students.

Retrieving the data :

By using 'Create table' command, the table has been created and 'insert into' command is used for insertion of data into the table. The data filled in the table can be viewed or displayed on the screen by "SELECT" command.

To view the data contents in the table, the 'SELECT' command is used.

Sytanx : SELECT * FROM Tablename;

Example : SQL > SELECT * FROM emp;

This prints the complete data in the table.

Example : SQL > SELECT * FROM STUDENT;

Result :

NAME	AGE	ADDRESS	CLASS
AMISHA	19	KANPUR	B.Sc 1st Year
TANU	18	NAGPUR	B.A.
YUVRAJ	10	VARANASI	6
AKANSHA	12	MIRZAPUR	8

This shows the values in the table student.

Retrieving data of selected columns and all rows

Syntax : SELECT <columnnames> FROM <tablename>;

Example : SQL > SELECT ename, desig FROM emp;

Output :

ENAME	DESIG.
KALPANA	CEO
AMISHA	Manager
NOOPUR	Manager
RAJEEV	Admin Officer.

It will display the name of the employees and their designation for all the records.

Retrieing data of selectred columns and selected rows :

Syntax : SELECT <columnnames> FROM <tablename> [WHERE Condition];

Example : SQL > SELECT ename, desig FROM emp WHERE dest.no=10;

The command will give the name and designation of all the exployees from department number 10. The WHERE clause expects three element : a column name, a comparison operator and content or the List values to be compared against the column values from the table.

Sorting Data in a Table :

The ORDER BY Command is used for sorting the data in either increasing or decreasing order, depending in the condition specified in the select statement.

Syntax : SELECT * FROM <statement> ORDER BY <columnname> [Sort order]

Example : SQL > SELECT * FROM emp ORDER BY sal;

It will give the data in the increasing order of salary, all records having null values in the salary column will be displayed in the last.

Output :

Emp no.	Ename	Sal.
1002	RADHA	8,000/-
1001	RAVI	9,200/-
1006	RAM	10,800/-
1008	RAJESH	11,000/-

For getting the data in decreasing order of salary, we use DESC at the end.

EXAMPLE : SQL > SELECT * FROM emp ORDER BY Sal DESC;

Updating the rows

Update Command is used to change the values of one or more columns in a row or a set of rows.

Syntax : UPDATE <tablename> SET <column name> = <expression> [WHERE <search condition>]

Example : SQL > UPDATE emp Set ename = 'Akash' WHERE empno = 1002;

Example : SQL > UPDATE emp SET Sal = (SELECT MAX (sal) from emp)

WHERE ename = 'Akash';

This command will update the salary of Akash to the Maximum salary of emp table.

Deleting the row

To remove one or more record from a table, we use Delete command.

Syntax : DELETE FROM <tablename> [WHERE Condition];

Example : SQL > DELETE FROM emp WHERE ename = 'Kalpana';

DCL (Data Control Language)

It is the component of SQL statement that control access to data and to the database. Occasionally DCL statements are grouped with DML statements,

Through DCL, we can

1. *Commit*: Save work done.
2. *Roll back*: Restore database to original since the last commit.

Fundamental Commands of DCL :

When an INSERT, UPDATE and DELETE operations are performed on the database, we can reverse or rollback the work what we have done. We process of saving the work done permanently or routing back work is controlled by Two commands namely: COMMIT and ROLLBACK.

COMMIT make changes in the Current transaction pemanent and ends the transaction.

Syntax : COMMIT;

Example : SQL > COMMIT';

Output : Commit Complete

Syntax : ROLLBACK;

Example : SQL > ROLLBACK;

Output : ROLLBACK Complete.

INSERT, UPDATE & DELETE Commands are not made final untill we commit them, All the DDl commands have an implicit commit command to execute and saves all the DML operations on the databse pemanently. On the other hand ROLLBACK Command is used to undo work. But it only undoes those operations, which are not committed to the server permanently. It searches the last commit executed and undo all the operations after that.

Data Query Language (DQL) :

It is the component of SQL statement that allows getting data from the database and imposing ordering upon it. It includes the select statement. This command is the heart of SQL. "*A query is a statement requesting the retrieval of information.*" Through DQL, we can

1. **Select**: Retrieve data from the database.

1.14. OVERVIEW OF HIERARCHICAL, NETWORK AND RELATIONAL DBMS

In hierarchical DBMS, data is organised in a series like a family tree or organisation chart. As with a family tree, the hierarchical database has branches made upof parent and child records. Each parent record can have multiple child records, but a child record can have only one parent.

A network DBMS, is similar to hierarchical model except that each child record can have more than one parent. Hence, this model allows having one-to-one, one-to-many, and many-to-many relationships.

A relational DBMS represents all data in the database as simple two-dimensional tables called relations that are the logical equivalent of files. The tables in relational databases, are organised in rows and columns, simplifying data access, and manipulation.

Summary

- A database management system is a collection of interrelated data and a set of programs to access those data.
- DBMS can also be define as in interface between the application program and the operating system to access or manipulate the database.
- Data modes can be define as a collection of conceptual tools for describing tools for describing data, data relationships, data semantics and consistence constraints. They are of five types : Relational data. Model, Object-oriented data model, Network data model.
- The collection of information stored in the database at a particular moments is called an instance of the database.
- The overall design of the database is called the database schema. It is of three types : Physical schema, conceptual schema and External schema.
- The ability to modify a schema definition in the next higher level is called Data independence. It is of two types : Physical data independence and Logical data independence.
- Data definition language (DDL) is used to create, modify and delete database.
- Data manipulation language (DML) is used to access or manipulate database. Here we can update, Add, insert and modify the database. It is of two types : Procedural DML and Non-procedural DML.
- Data control language (DCL) is used to control the access to data and to database. Here, we can commit and Rollback.
- Data query language (DQL) is used for getting data from the database. Though DQL, we can select data from the database.

Selected Exercises

1. What do you mean by DBMS? Also mention its applications.
2. What are the main differences between a file processing system and a DBMS?
3. What is the difference between logical data independence and physical data independence?
4. What do you mean by External schema, Conceptual schema and Internal schema?
5. What are interfaces? Explain in brief.
6. What do you mean by Data Abstraction ? Explain its types in details.
7. Explain following terms : Data, Database, Database System.
8. What are Data Models? Also explain their types.
9. Explain DDL, DML and DCL in detail?
10. What is the difference between procedural and non-procedural DML?
11. Discuss the three level architecture of database system.

❑❑❑

Chapter

2 Data Modelling Using the E-R Model

Inside this Chapter

2.1. E-R MODEL CONCEPTS

The entity-relationship (E-R) model is a high-level data model. It is based on a perception of a real world that consists of a collection of basic objects, called *entities*, and of relationships among these objects. It was developed to facilitate database design by allowing specification of an enterprise schema, which represent the overall logical structure of a database.

Entity. An entity is an object that has its existence in the real world. It includes all those "things" about which data is collected. An entity may be a tangible object such as a student, a place or a part. It may also be non-tangible such as an event, a job title or a customer account. For example, if we say that a customer buys goods, it means customer and goods are entities.

Diagrammatically, entities are represented in rectangles.

CUSTOMER		GOODS

Entities

An Entity Set : It is a set of entities of the same type that share the same properties, or attributes. The set of all persons who are customers at a given bank, example, can be defined as the entity set customer.

Attributes : Attributes are units that describe the characteristics or properties of entities. In a database, entities are represented by tables and attributes by columns. For example, a customer entity might have numerous attributes such as code, name and addresses. Similarly, the goods entity may have attributes like code and price. They are drawn in elliptical shapes along with the entity rectangles.

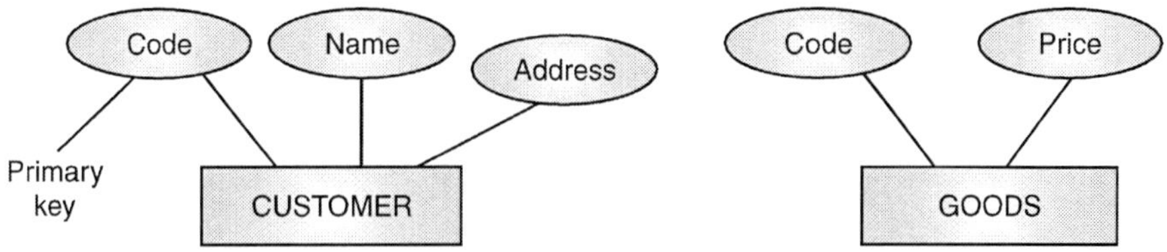

Fig. 2.1. Attributes

Domain : The domain of an attribute is the collection of all possible values, an attribute can have.

Field/Column : A column represents one related part of a table and is the smallest logical structure of storage in a database. It holds one piece of information about on item of subject.

It is generally used for a group of alphanumeric characters.

Record/Row : A record is a collection of multiple related fields that can be treated as a unit. In database terminology, each row is a record. However database experts prefer to use the word row, since record sometimes has other connotations outside of relational databases.

Table : A table is a named collection of logically related multiple records. Depending on the database software, a table can be referred to as a file.

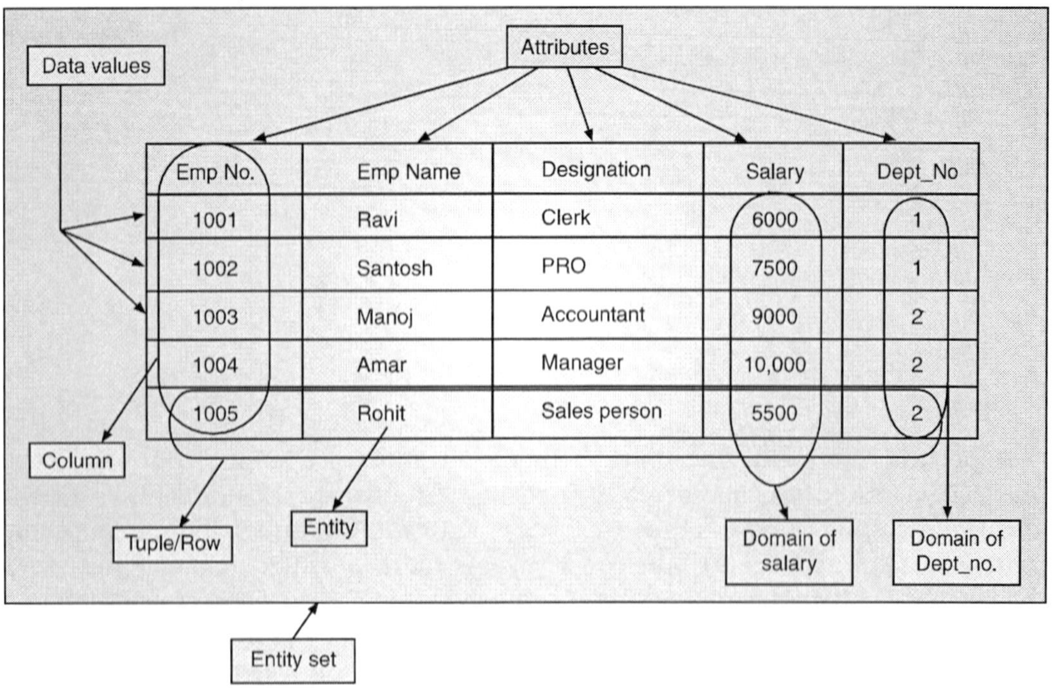

Fig. 2.2.

2.2. NOTATION OF E-R DIAGRAM *(UPTU, 2004-05)(IMP.)*

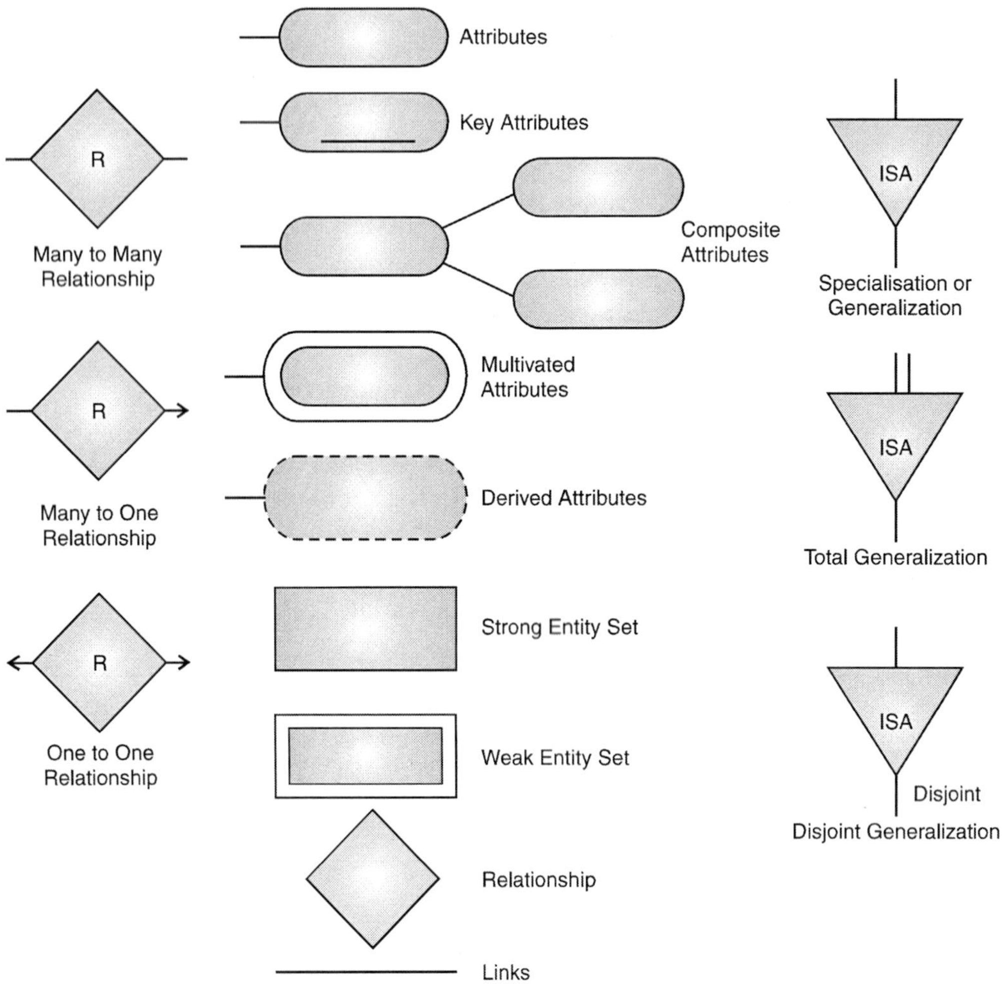

Fig. 2.3.

Simple and Composite Attributes : Simple attributes are those attributes which can not be divided into subparts. Whereas composite attributes are those attributes which can be divided into subparts (*i.e.*, other attributes). For example, an attribute name could be considered as a composite attribute consisting of first name, middle name and last name. Using composite attributes in a design schema is a good choice if a user will wish to refer to an entire attributes on some occasions, and to only a component of the attribute on other occasions. Suppose we were to substitute for the sustomer entity-set attributed customer-street and customer-city the composite attribute address with the attributes, street, city, state and zip-code. Composite attributes help us to group together related attributes, making the modeling cleaner.

Composite attribute may also appear as a hierarchy. In the composite attribute address, its component attribute street can be further divided into street-number, street-name, and apartment-number.

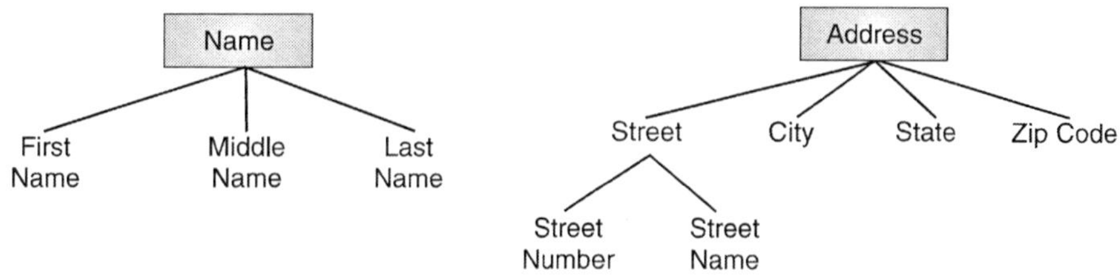

Fig. 2.4. Compositye Attributes.

Single-Values and Multvalued Attributes : The attributes which has a single-value for a particulary entity is called single-valued attributes, whereas attributes which has a set of values for a specific entity is called multivalued attributed. For example, the Loan number attributes for a specific loan entity refers to only one loan number, such type of attributes are single-valued whereas phone number attributes, where the employess may have zero, one or several phone numbers. This type of attributes is said to be multivalued.

Derived Attributes : The value for this type of attribute can be derived from the values of other related attributes or entities. For example, suppose that the customer entity set has an attribute age, which indicates the customer's age. If the customer entity set also has an attibute date-of-birth, we can calculate age from date-of-birth and the current date. Thus, age is a derived attributes. In this case Date-of-birth may be referred to as a base attribute, or a stored attribute. The value of a derived attribute is not stored, but is computed when required.

An attribute takes a null value when an entity does not have a value for it. The null value may indicate "not applicable".

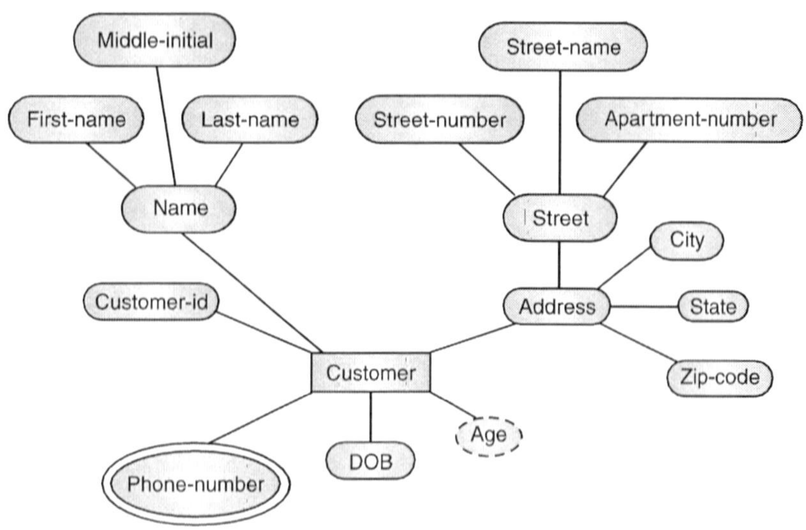

Fig. 2.5. E-R diagram with composite, multivalued and derived attributes.

2.3. EXAMPLES OF E-R DIAGRAM

Example 2.1.

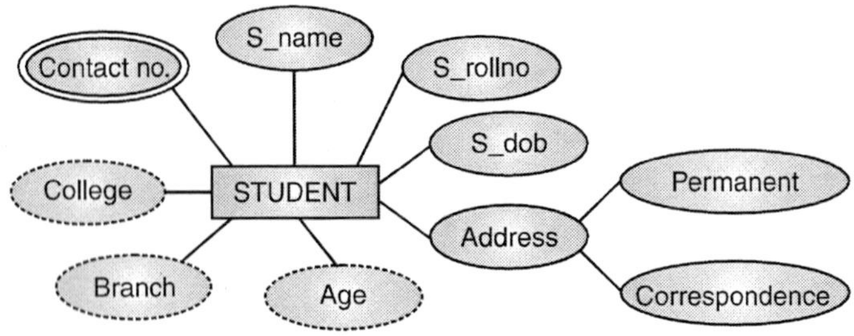

Fig. 2.6. ER diagram for student

Example 2.2. ER-diagram for hospital with a set of patients and medical doctors.

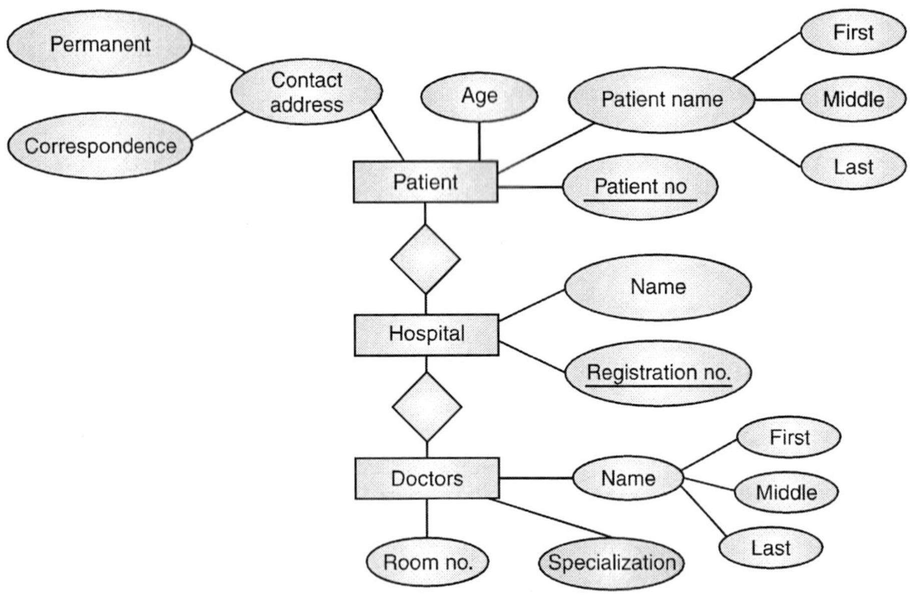

Fig. 2.7. ER diagram for hospital

Example 2.3.

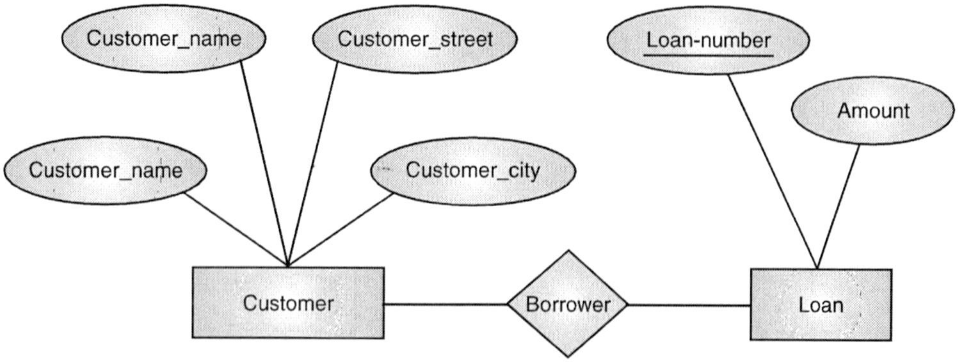

Fig. 2.8. ER diagram corresponding to customer and loans

Example 2.4.

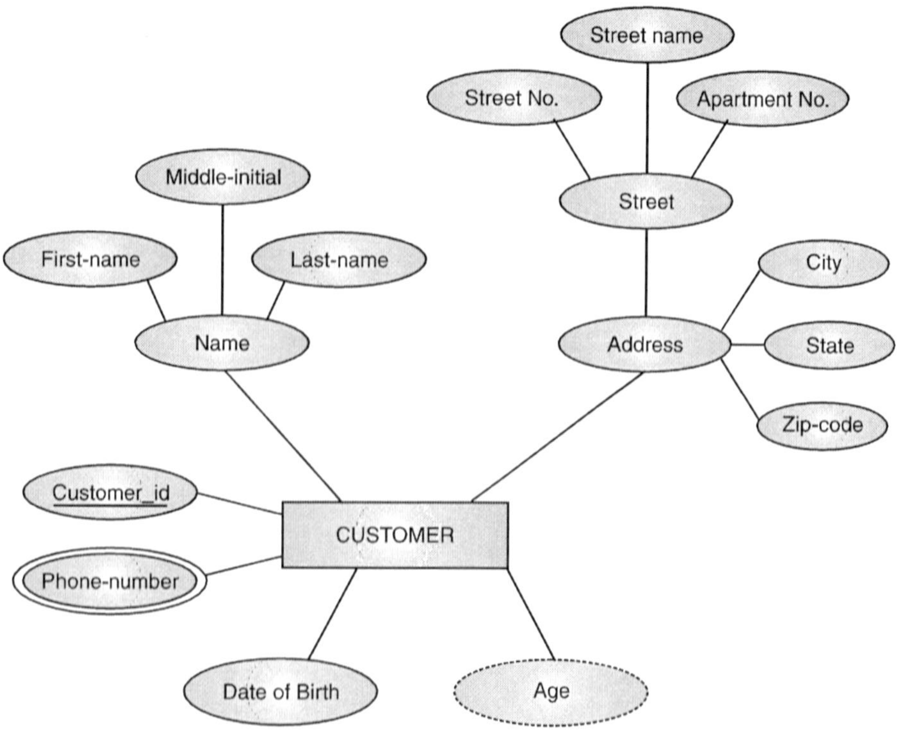

Fig. 2.9. ER diagram with composite, multivalued and derived attributes

Example 2.5. E-R diagram for Airline Reservation System

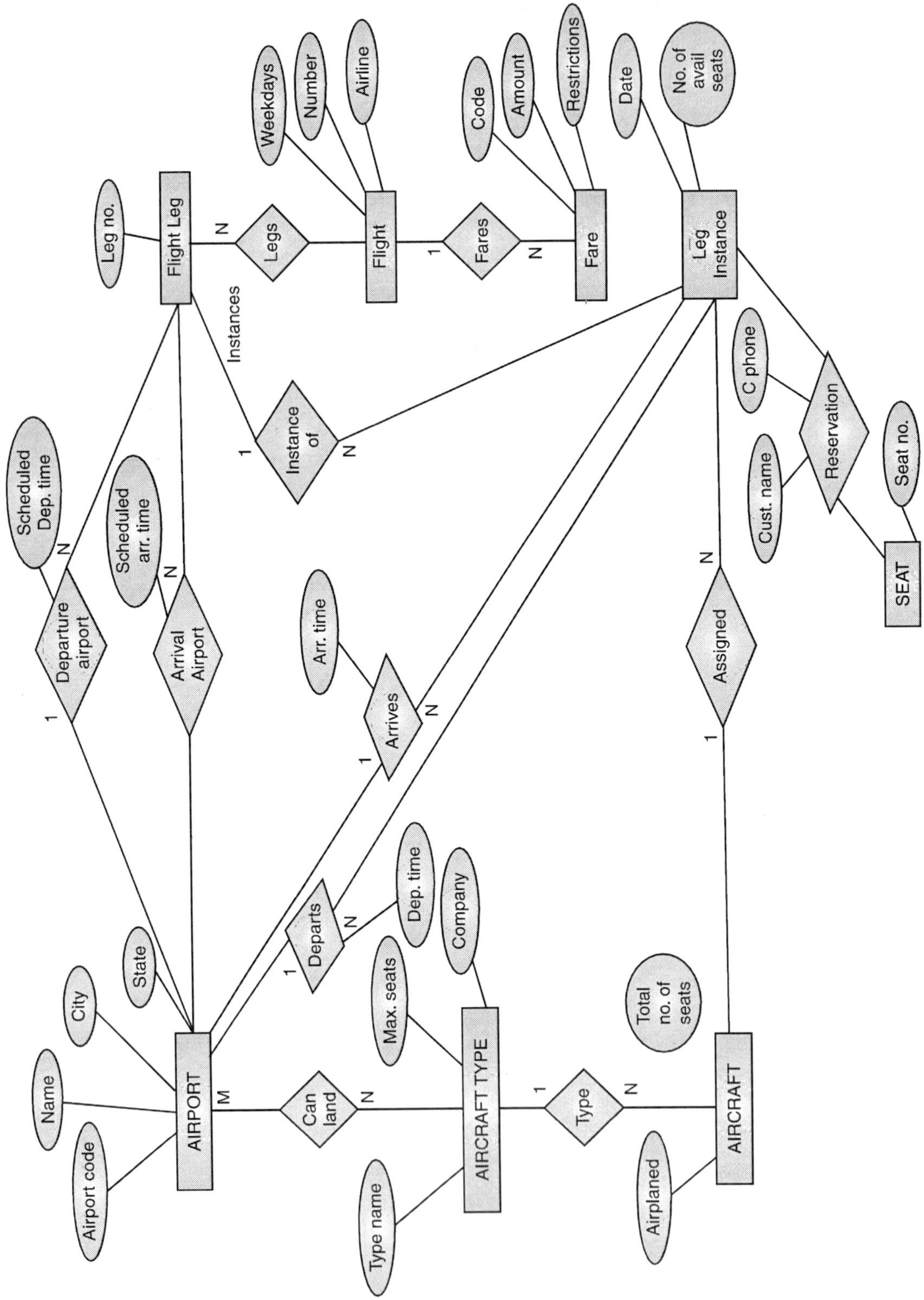

Fig. 2.10.

Example 2.6. **ER diagram of banking system.**

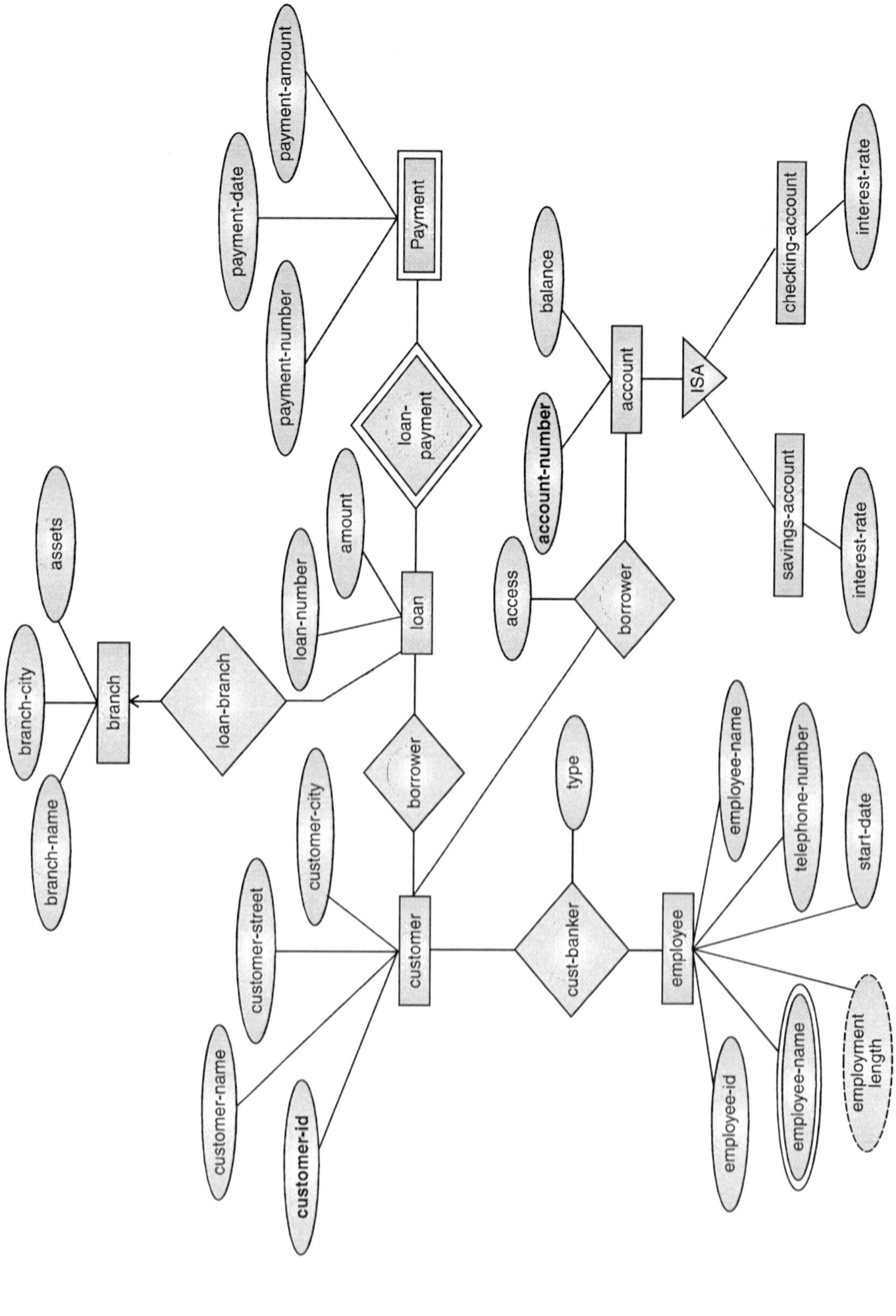

Fig. 2.11.

Example 2.7. Give an entity-relationship diagram for a database showing fatherhood, motherhood and spouse relationship among men and women.

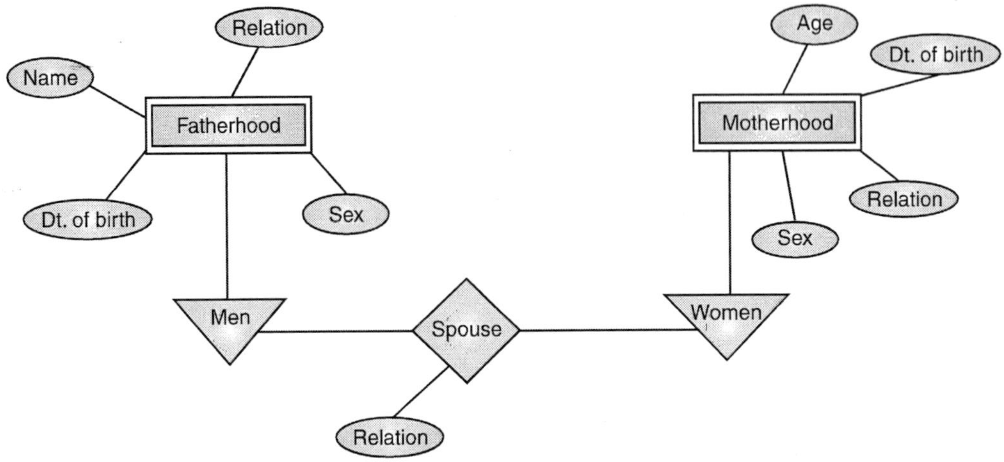

Fig. 2.12.

2.4. WEAK-STRONG ENTITY SET

An entity set that does not possess sufficient attributes to form a primary key is called a weak-entity set. One that does have a primary key is called a strong entity set. A weak-entity set does not have a primary key, but we need a means of distinguishing among the entities. The distinction of a weak-entity set is a set of attributes that allows this distinction to be made.

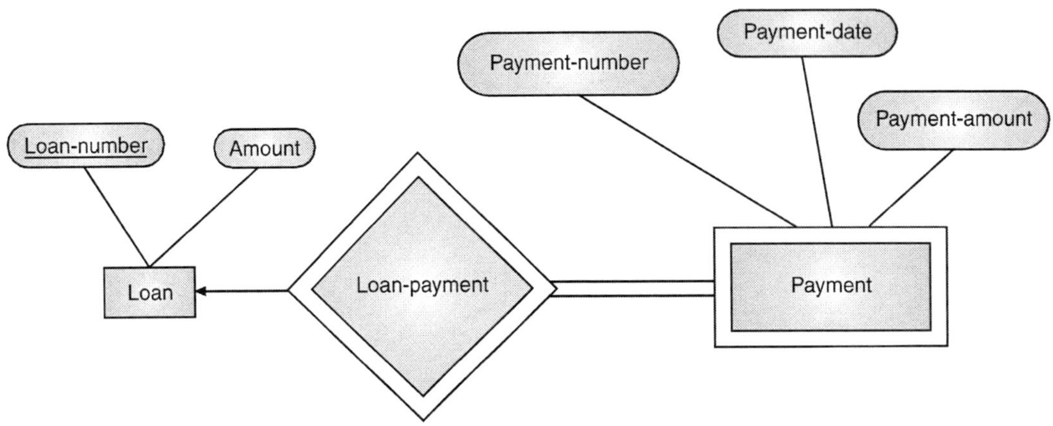

Fig. 2.13. E-R diagram with a weak entity set.

Example 2.8. We can convert any weak entity set to a strong entity set by adding appropriate attribute. Why, then do we have weak entity sets?

Solution : An entity set is called an weak entity set if its existence depends on other entities (called *strong entities*). Thus a week entity set does not have sufficient attributes to form a primary key.

An entity set which have a primary key is termed as a strong entity set.

An weak entity set can be convert into strong entity set with the help of depending strong entity set, and the primary key will be the union of primary key and determinant attributes.

The entity set departure which has only attribute date is a weak entity set. This is because if each departure entity is distinct, different flights may have departure on same day, so different flights may share the same data value.

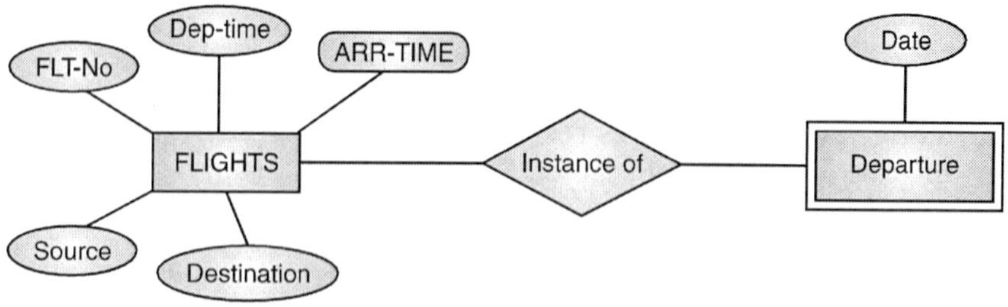

Fig. 2.14.

2.5. CARDINALITY OF RELATIONSHIP

Relationship : Relationship is an association, dependency or unit between two or more entities and is represented by a diamond symbol. All relations have three components :

1. **Name :** It is the title or entity identifier such as goods bought.
2. **Degree :** It represents the number of attributes (field) associated with the table or relation.
3. **Cardinality :** It can be thought of as the maximum number of records, in one entity that are linked to a single zero in another entity and vice-versa. Tuples are the number of records in a relation.

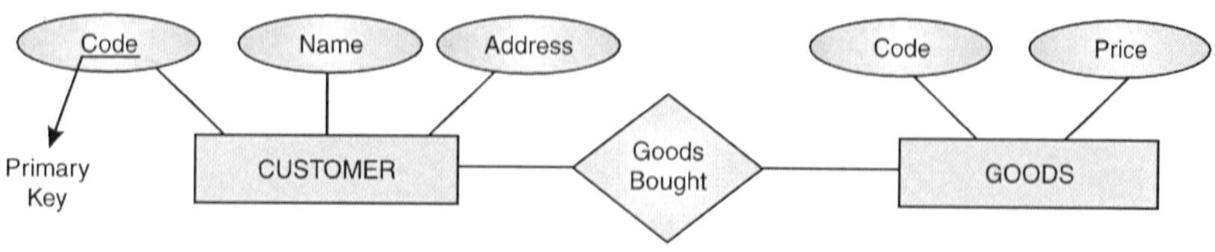

Fig. 2.15. Entities, Attributes and Relationship

Even though a relationship may involve more than two entities, the most commonly encountered relationships are binary, involving exactly two entities. Generally, such binary relationships are of three types : one-to-one, one-to-many and many-to-many.

One-to-one Relationship (1 : 1)

In one-to-one relationship, one record in a table is related to only one record in another table. For *e.g.* a department cannot be headed by more than one department head.

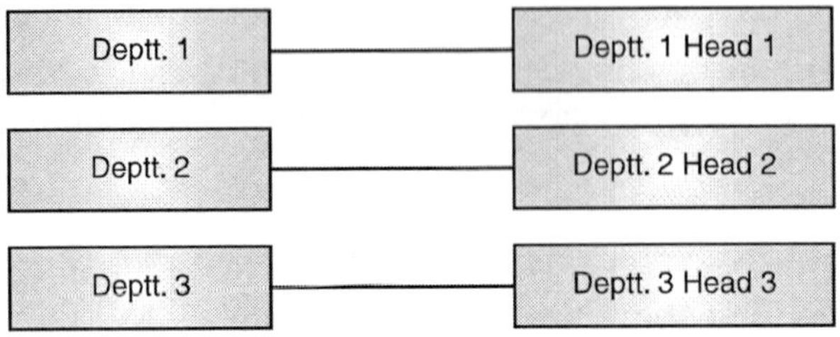

Fig. 2.16. One-to-one Relationship

One-to-Many Relationship (1 : M)

In one-to-many relationship, one record in a table (parent table) can be related to many records in another table (child table). For example, a father may have more than one child but the child has only one father.

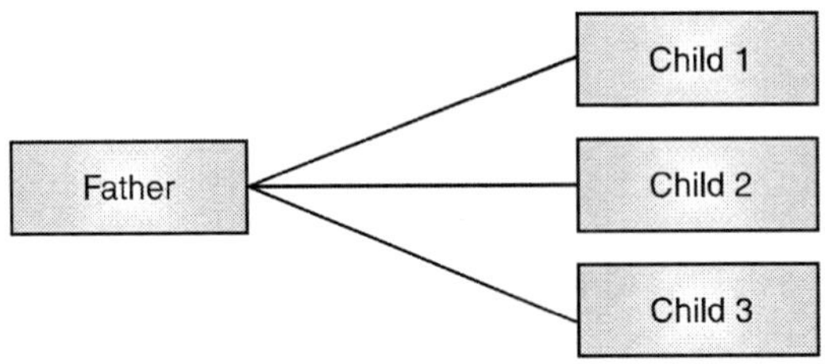

Fig. 2.17. One-to-Many Relationship

Many-to-Many Relationship (M : M)

In many-to-many relationship, one record in a table can be related to one or more records in another table, and one or more records in the second table can be related to one or more records in the first table. For example, a customer can buy many goods and many customers can buy the same goods.

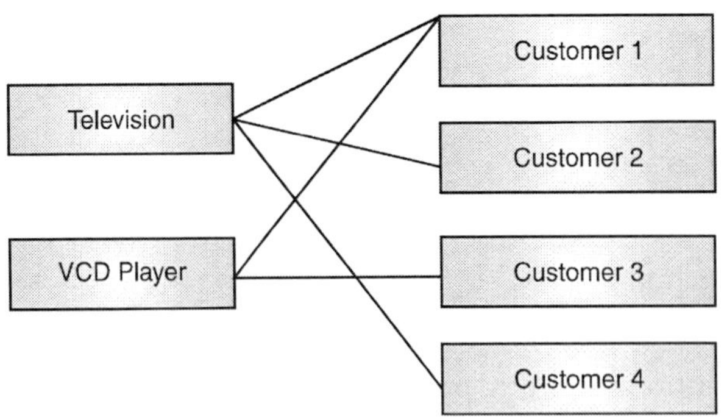

Fig. 2.18. Many-to-many Relationship

If there are two entity types participating in a relation then it is called a relationship type of degree two (binary), and a relationship type is called degree three (ternary) if there are three entity types participating. If there is a relationship that involves *N* entities is called *N-ary relationship*.

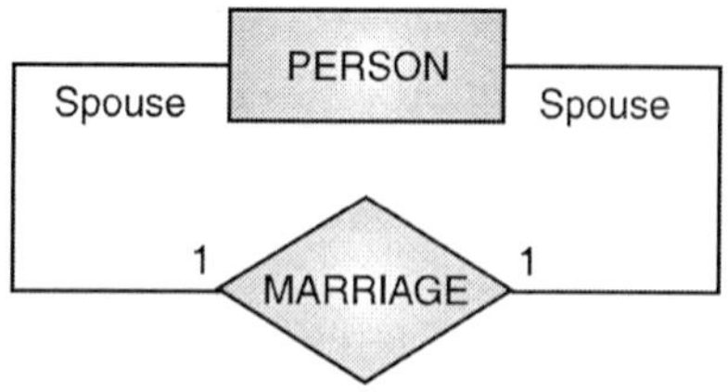

Fig. 2.19.

2.6. EXTENDED E-R MODEL

The Extend Entity Relationship Model (EER-Model) includes all the modelling concepts of the E-R Model and the concepts of specialization, generation higher level and lower level entity set, Attributes Inheritance and Aggregation.

There are basically four conceptys of EER-Model :

- Specialization
- Generalization
- Attribute Inheritance
- Aggregation

2.6.1. Specialization

The process of designating subgroupings within an entity set is called specialization for example, an entity set person, with attributes name, street and city. A person may be further classified as one of the following :

- Customer
- Employee

Each of these person types is described by a set of attributes that includes all the attributes of entity set person and possibly additional attrbutes, for example, customer entities may be described further by the attribute customer_id, whereas employee entities may be described further by the attributes employee_id and salary. The specialization of person allows us to distinguish among persons according to whether they are employees or customers.

Consider another example, suppose the bank wishes to divide accounts into two categories, cheking account and savings account—savings accounts need a minimum balance, but the bank may set interest rates differently for different customers, offering better rates to favoured customers.

The bank could then create two specializations of account, namely saving account and checking account.

We can apply specialization repeatedly to refine a design scheme. For instance, bank employees may be further classified as one of the following :

- Officer
- Teller
- Secretary

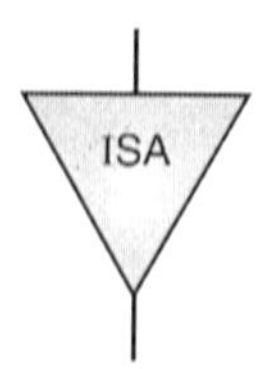

Fig. 2.20.

Each or these employee types is described by a set of attributes that includes all the attributes of entity set employee and additional attributes. For example, officer entities may be described further by the attribute office number, teller entities by the attributes station number and hours-per-week and secretary entities by the attribute hours-per-week.

In terms of an E-R diagram, specialization is depicted by a triangle component labelled ISA.

The label ISA stands for "is a" and represents, for example, that a customer "is a" peson. The ISA relationship may alo be referred to as a superclass-subclass relationship. Higher and lower-level entity sets are depicted as regular entity set, *i.e.*, as rectangular containing the name of the entity set.

2.6.2. Generalization

The refinement from an initial entity set into successive levels of entity subgrouping represents a top-down design process in which distrinctions are made explicit. The design process may also process in a bottom-up manner, in which multiple entity sets are synthesized into a higher level entity set on the basis of common features. The database designer may have first identified a customer entity set with the attributes name, street, city and customer_id and an employee entity set with the attributes name, street, city, employee_id and salary. These are similarities between the customer entity set and the employee entity set in the sense that they have sevral attributes in common. This commonality can be expressed by generalization, which is a containment relationship that exists between a high-level entity set and one or more lower-level sets. In other example, person is the higher-level entity set and customer and employee are lower-level entity sets. Higher- and lower-level entity sets also may be designated by the terms super class and subclass, respectively. The person entity set is the superclass of the customer and employee subclass.

"Generalization is a simple. Inversion of specialization".

Differences in the two approaches, *i.e.*, Specialization and Generalization may be characterized by their starting point and overall goal. Specialization stems from a single entity set, it emphasises differences among entities within the set by creating district lower-level entity sets. These lower-level entity sets may have attributes, or may participate in relationships, that do not apply to all the entities in the higher-level entity set. The reason to apply specialization is to represent such distinctive features. If customer and Employee neither have attributes that person entities do not have nor participate in different relationships than those in which person entities participate, there would be no need to specialize the person entity set.

Generalization proceeds from the recognition that a number of entity sets share some common features. On the basis of their commonalities, generalization synthesizes these entity sets into a single, higher-level entity set. Generalization is used to emphasize the similarities among lower-level entity sets and to hide the differences; it also permits an economy of representation in that shared attributes are not repeated.

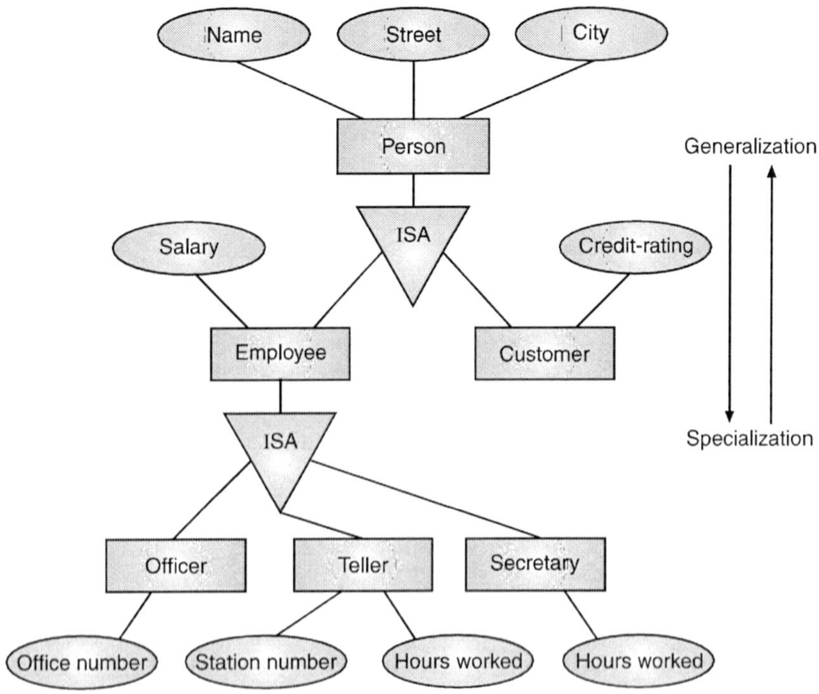

Fig. 2.21. E-R diagram with Generalization and Specialization

2.6.3. Attribute Inheritance

A crucial property of the higher- and lower-level entities created by specialization and generalization is attribute inheritance. The attributes of the higher-level entity sets are said to be inhertied by the lower-level entity sets. For example, customer and employee inherit the attributes of person. Thus, customer is described by its name, street and city attributes and additionally a customer_id attribute; employee is described by its name, street and city attributes and additionally employee_id and salary attributes.

The previous diagram just above depicts a hierarchy of entity sets. In the figure, emplotyee is a lower-level entity set of the officer, teller and secretary entity sets. In a hierarchy, a given entity set may be involved as a lower-level entity set in only one ISA relationship, *i.e.*, entity sets in this diagram have only single inheritance. If an entity set is a lower-level entity set in more than one ISA relationship, then the entity set has multiple inheritance and the resulting structure is said to be a "Lattice".

2.6.4. Aggregation

Aggregation is an abstraction through which relationships are treated as hgiher level entities. Thus, for our example, we regard the relationship set works-on (relating the entity sets employee, branch and job) as a higher-level entity set called works-on. Such an entity set is treated in the same manner as is any other entity set. We can then create a binary relationship manages between works-on and manager to represent who managers what tasks.

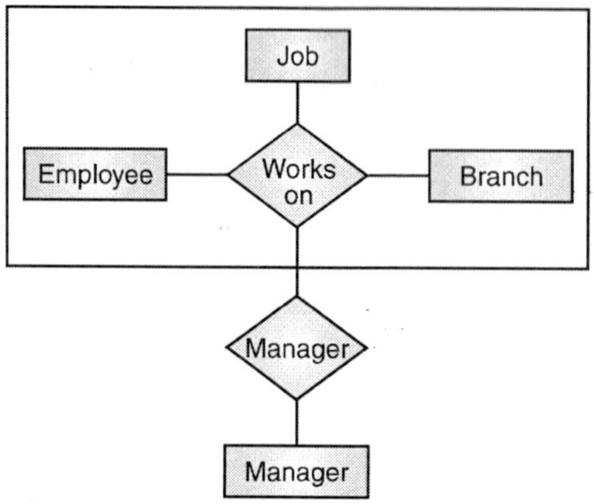

Fig. 2.22. E-R diagram with aggregation.

2.7. TRANSLATING YOUR E-R MODEL INTO RELATIONAL MODEL

Database can be represented with the help of E-R notations and these notations (diagrams) can be reduced to collection of tables. For every entity set, relationship set in a database can be represented in tabular form. The name of the table is the name of corresponding entity set of relationship name. Each table would have columns, the number of columns would be equal to the number of attributes each entity set or relationship set have.

For *example* :

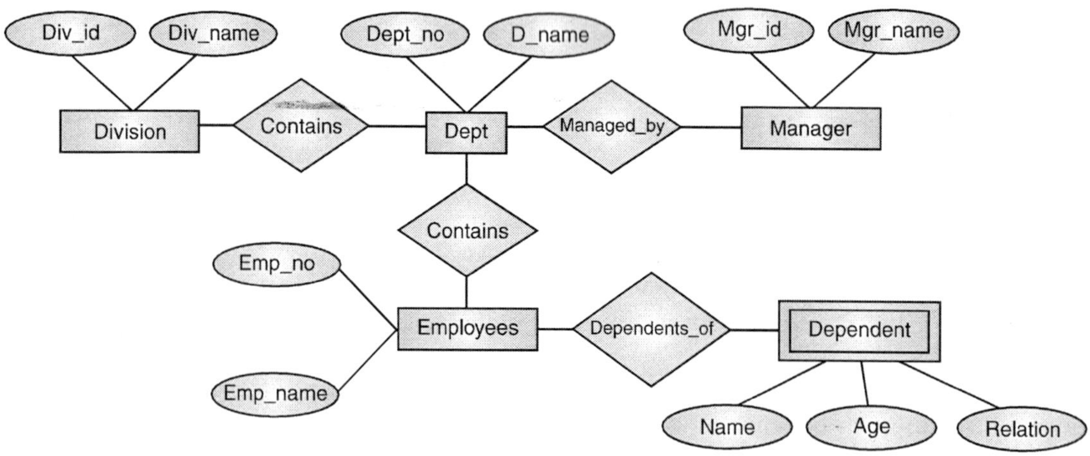

Fig. 2.23. E-R diagram

We shall convert E-R diagram into tables :

We divide this E-R diagrams into three sections :

1. Strong Entity Sets (Division, Dept, Manager, Employees).
2. Weak Entity sets (Dependents)
3. Relation sets (contains, managed by, Have Dependents of).

2.7.1. Reducing Strong Entity sets into tables

Consider the entity sets division of the E-R diagram. This entity set have two attributes Div-id and Div-name. We shall represent this entity set by a table called Division with two columns.

Div_id	Div_name
D1	Marketing division

Division Relation

Similarly, entity set 'Depth' has two id Dept_no and D_name, which will be represented by table Dept. Entity set manager also has two id Mgr_id and Mgr_name which will be represented by table manager and entity set employee has two id emp_no and Emp_name which will be represented by table employee.

2.7.2. Reducing Weak Entity Set into Table

In E-R diagram shown, we have only one weak entity set Dependents. This Entity set has three attributes Name, Age and Relationship.

2.7.3. Reducing Relationship Sets into Tables

In E-R diagram shown, we have four relationship sets contains, managed-by Have, Dependents, of contains relationship set involves two entity sets Division with primary key Div_id and Dept with primary key Dept_no.

Managed-by relationship set involves two entity sets Dept with primary key Dept_no and manager with primary key Mgr_id.

Have relationship set involves two entity sets Dept. with primary key dept_no and employee with primary key Emp_no.

Div_Id	Dept_no
D1	DEPT01
D2	DEPT02

Contains Relation

Dept_no	Mgr_id
DEPT01	MGO1
DEPT02	MG02

Managed-by Relation

Dept-no	D-name
DEPT01	New Business Development Department
DEPT02	Higher Education Department

Entity Dept. Relation

Magr_id	Mgr_name
MG01	Rajesh
MG02	Ravikant

Entity Employee Relation

EMP_NO	EMP_NAME
10001	Akshay
10002	Ashish
10003	Sparsh
10004	Meenal
10005	Tanu
10006	Rahul

Entity Manager Relation

Dept_no	EMP_no
DEPT01	1001
DEPT01	1002
DEPT01	1003
DEPT02	1004
DEPT02	1005
DEPT02	1006

Fig. 2.24. Have Relation

Example 2.9. Reduce the E-R Diagram into tables.

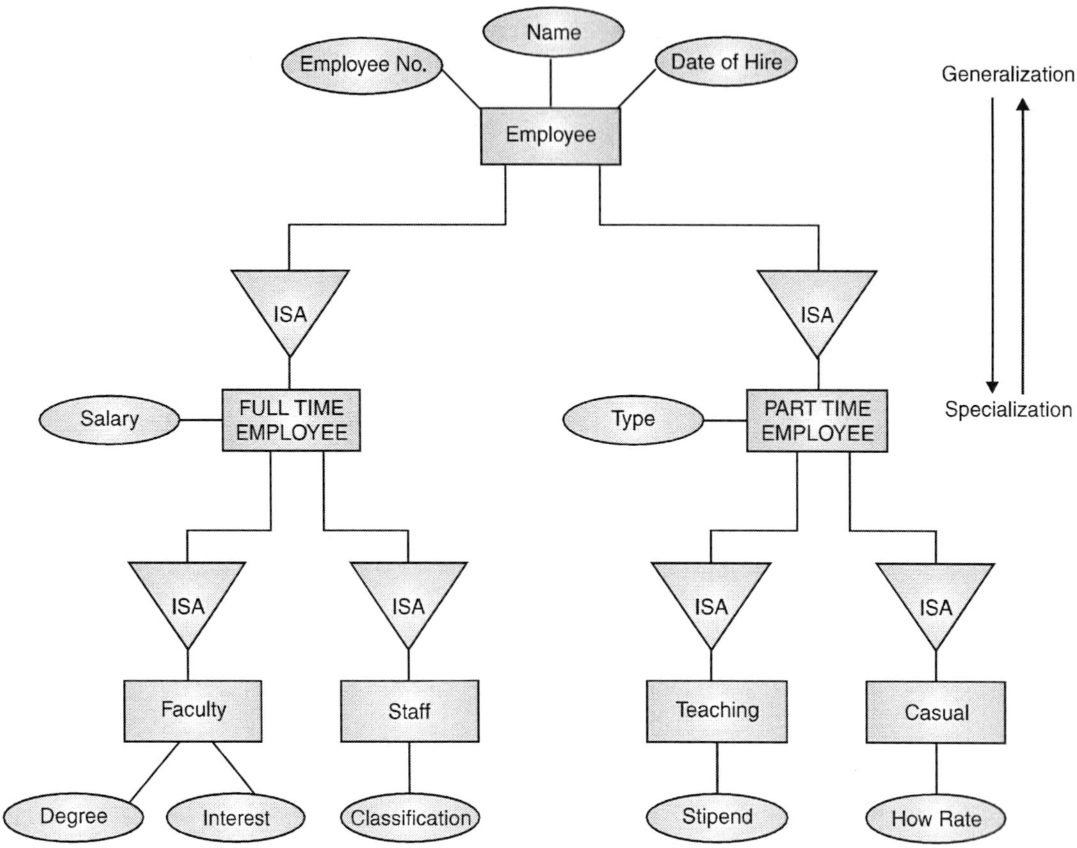

Fig. 2.25.

Methods :

1. First we identify different strong entity from the above E-R-diagram for example, employee, full time employee, part time employee, faculty, staff, teaching, casual.

   The number of table will be equals to the number of strong entity set.
2. Now identify the primary key in each entity.
3. Now create table of each entity alongwith their primary key.
4. Add the attributes of each entity with their entity.

Table 1. Employee

Emp. No	Name	Date of Hire
⋮	⋮	⋮
⋮	⋮	⋮
⋮	⋮	⋮

Table 2. Full Time

Emp. No.	Salary
⋮	⋮
⋮	⋮
⋮	⋮

Table 3. Part-Time

Emp. No	Type
⋮	⋮
⋮	⋮
⋮	⋮

Table 4. Faculty

Emp. No.	Degree	Interest
⋮	⋮	⋮
⋮	⋮	⋮
⋮	⋮	⋮

Table 5. Staff

Emp. No	Classification
⋮	⋮
⋮	⋮
⋮	⋮

Table 6. Teaching

Emp no.	Stipend
⋮	⋮
⋮	⋮
⋮	⋮

Table 7. Casual

Emp. No	Half-rate
⋮	⋮
⋮	⋮
⋮	⋮

SOLVED EXAMPLES

Example 2.10. **Construct an E-R for a car insurance company whose customers own one-or-more cars each. Each car has associated with it zero to any number of recorded accidents.**

Solution :

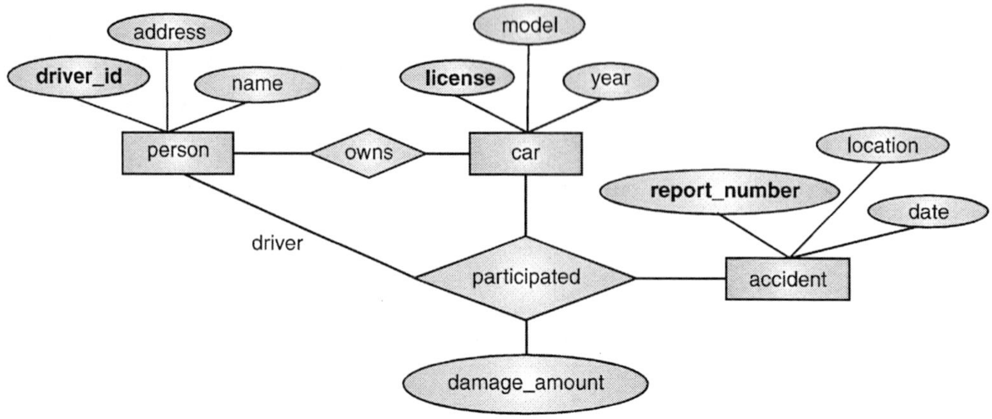

Fig. 2.25. (a) E-R diagram for a car insurance company

Example 2.11. Construct an E-R diagram for university registrar's office

Solution : Assumptions made are

(*i*) A class meets only at one particular places time. This E-R diagram cannot model a class meeting at different places at different times. (*ii*) There is no guarantee that the database does not have two classes meeting at the same place time.

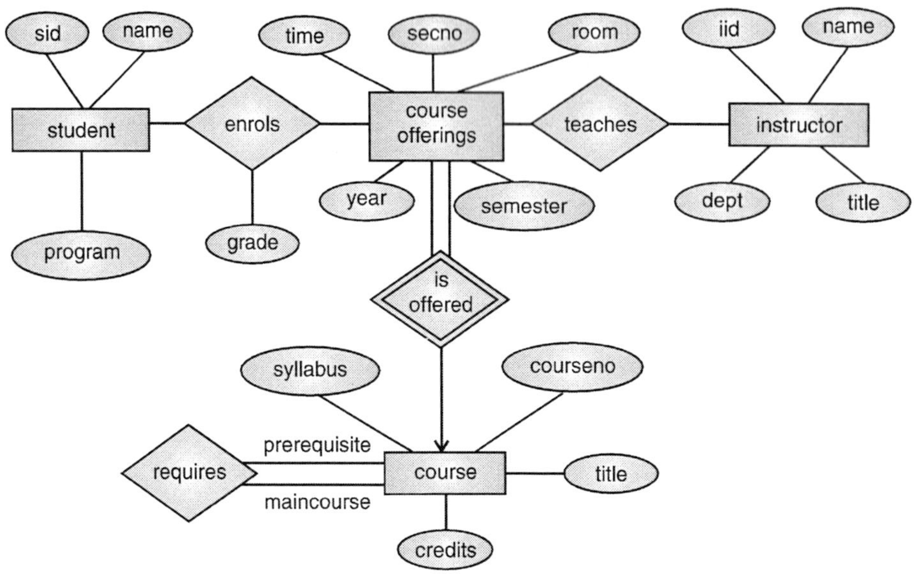

Fig. 2.25. (b) E-R diagram for a university

Example 2.12. Construct an E-R diagram that models exams as entities and users a ternary relationship, for the database.

Solution :

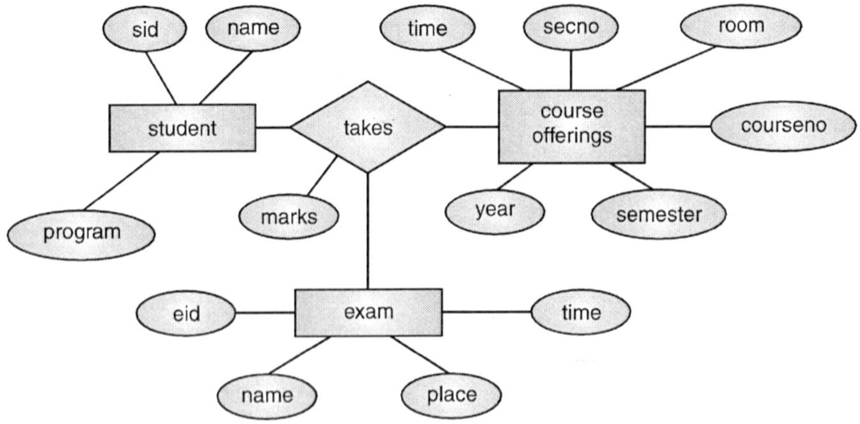

Fig. 2.26. E-R diagram for marks database

Example 2.13. Construct an E-R diagram that used only a binary relationship between students and course-offerings. Make sure that only one relationship exists between a particular student and course-offering pair. Yet you can represents the marks that a student gets in different exams of a course offering.

Solution :

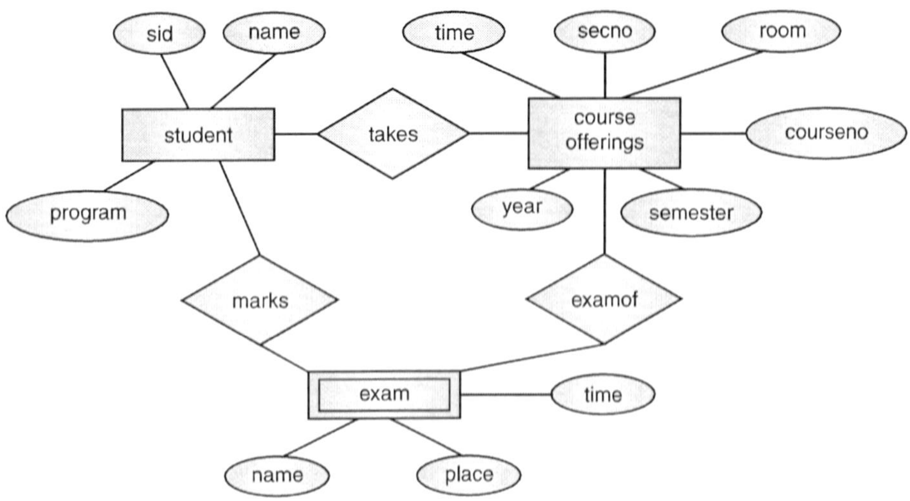

Fig. 2.27. E-R diagram for marks database

Example 2.14. Design an E-R diagram for keeping track of the exploits of your favourite sports team. You should store the matches played the scores in each match, the players in each match, and individual player statistics for each match. Summary satistics should be modelled as derived attributes.

Solution :

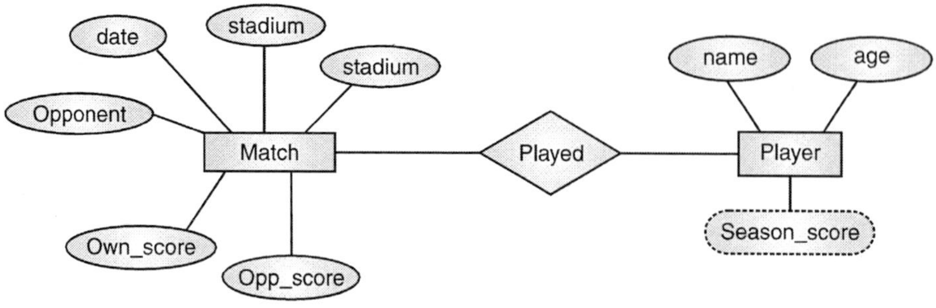

Fig. 2.28. E-R diagram for favourite team statistics

Example 2.15. Construct an E-R diagram in which the same entity set appears several times.

Solution : By using one entity set many times we are missing relationship in the model. For example, in the E-R diagram the students taking classes are the same students who are athletes, but this model will not show that.

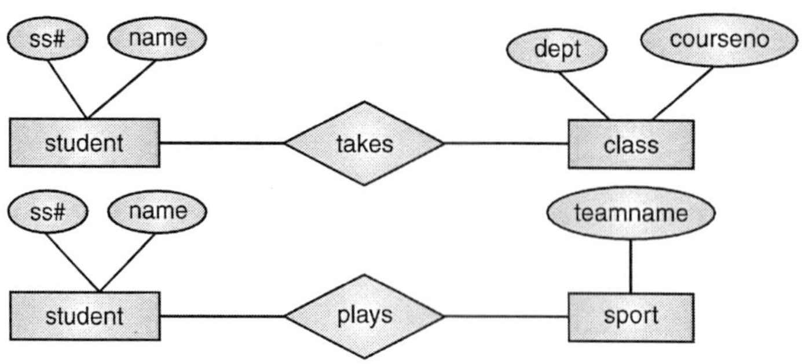

Fig. 2.29. E-R diagram with entity duplication

Example 2.16. Consider a university database for the scheduling of classrooms for final exams. This database could be modelled as the single entity set exam, with attributes course_name, section_number, room_number and time. Alternatively, one or more additional entity sets could be defined along with relationship sets to replace some of the attributes of the exam entity set, as

- **Course with attributes name, department, and c-number.**
- **Section with attributes s-number and enrollment, and dependent as a weak entity set on course.**
- **room-with attributes r-number, capacity and building.**

(a) Show an E-R diagram illustrating the use of all three additional entity sets listed ?

(b) Explain what application characteristics would influence a decision to include or not to include each of the additional entity sets.

Solution : (*a*)

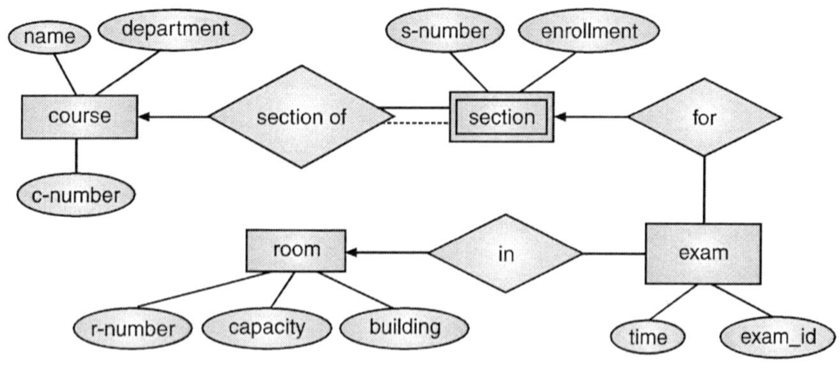

Fig. 2.30. E-R diagram for exam scheduling

(*b*) The additional entity sets are useful if we wish to store their attributes as part of the database. For the *course* entity set, we have chosen to include three attributes. If only the primary key (*c_number*) were included, and if courses have only one section, then it would be appropriate to replace the *course* (and *section*) entity sets by an attribute (*c_number*) of *exam*. The reason it is undesirable to have multiple attributes of course as attributes of exam is that is would then be difficult to maintain data on the courses, particularly if a couse has no exam or several exams. Similar remarks apply to the *room* entity set.

Example 2.17. When designing an E-R diagram for a particular enterprise, you have several alternative from which to choose ?

(*a*) What criteria should you consider in making the appropriate choice ?

(*b*) Design, three alternatives E-R diagrams to represent the university registrar's office. List the merits of each.

Solution : (*a*) The criteria to use are intuitive design, accurate expression of the real-world concept and efficiency. A model which clearly outlines the objects and relationship in an intuitive manner is better than one which does not, because it is easier to use and easier to change. Deciding between an attribute and an entity set to represent an object, and deciding between an entity set and relationship set, influence the accuracy with which the real-world concept is expressed. If the right design choice is not made, inconsistency and /or loss of information will result. A model which can be implemented in an efficient manner is to be preferred for obvious reasons.

(*b*) Consider three different alternatives for the problem.

- See figure 2.31
- See figure 2.32
- See figure 2.33

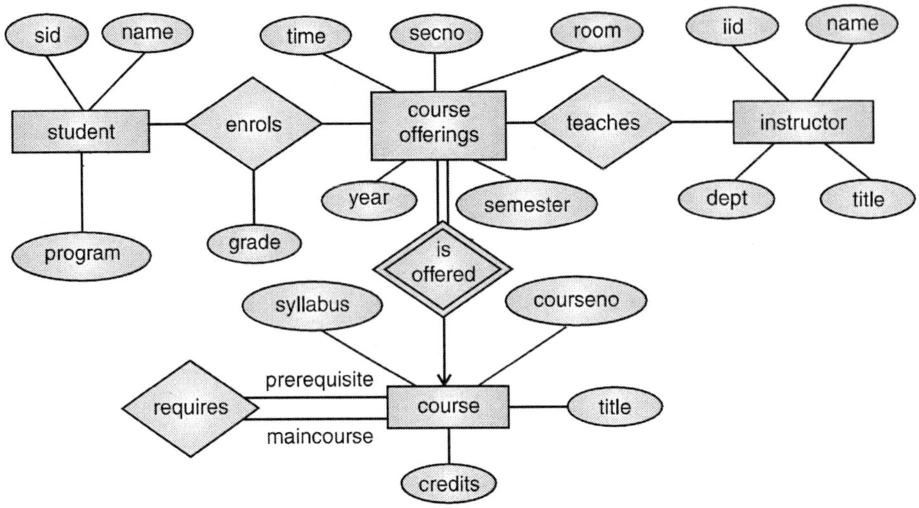

Fig. 2.31. E-R diagram for university.

Each alternative has merits, depending on the intended use of the database. However, it will be difficult to store all prerequisites (being a multi-valued attribute). Scheme treats prerequisites as well as classrooms as separate entities, making it useful for gathering data about perquisites and room usage. Since a registrar's office probably has to answer general questions about the number of classes a student is takin of what are all the prerequisites of a course, or where a specific class meets.

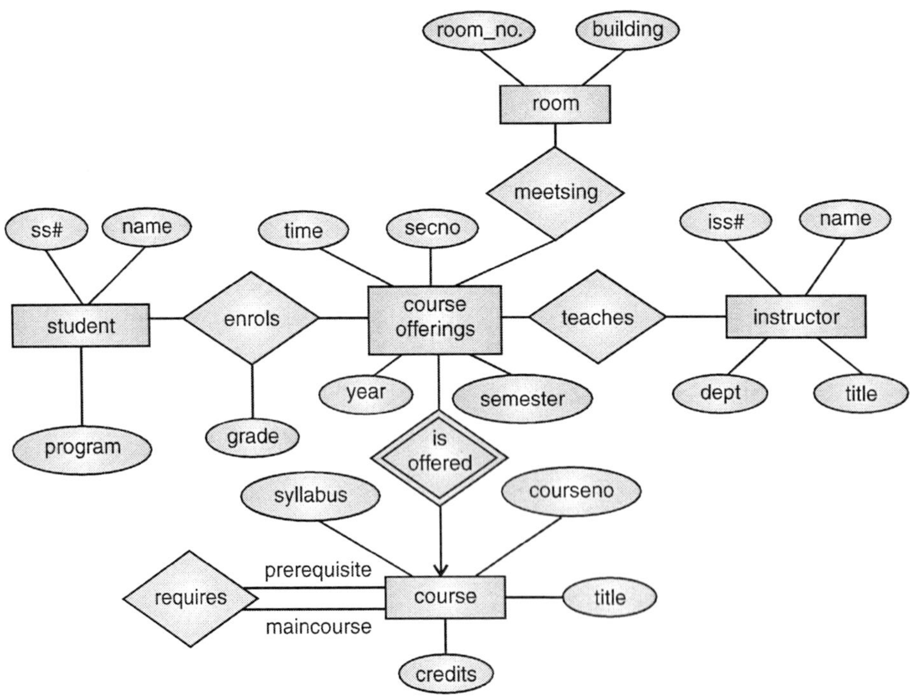

Fig. 2.32. E-R diagram for university

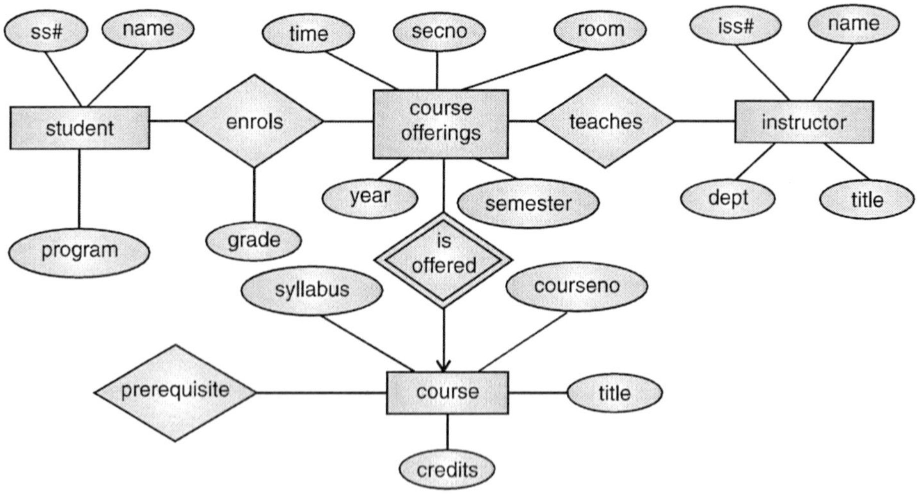

Fig. 2.33. E-R diagram for university.

Example 2.18. An E-R diagram can be viewed as a graph. What do the following mean in terms of the structure of an enterprise schemal

(*a*) The graph is disconnected.

(*b*) The graph is acyclic. *(UPTU, 2008-09)*

Solution : (*a*) If a pair of entity sets are connected by a path in an E-R diagram, the entity sets are related, though perhaps indirectly. A disconnected graph implies that there are pairs of entity sets that are urelated to each other. If we split the graph into connected components, we have in effect a separate database corresponding to each connected component.

(*b*) As indicated in the answer to the previous part, a path in the graph between a pair of entity sets indicates a (possibly indirect) relationship between the two entity sets. If there is a cycle in the graph then every pair of entity sets on the cycle are related to each other in at least two distinct ways. If the E-R diagram is acyclic then there is a unique path between every pair of entity sets and, thus, a unique relationship between every pair of entity sets.

Example 2.19. Suppose we informally derive the data in the database of a department store as follows :

(*i*) Each employee is represented. The data about an employee are his employee number, name, address and the department the works for :

(*ii*) Each department is represented. The data about a department are its name, employess, manager and items sold.

(*iii*) Each item solid is represented. The data about an item are its name, manufacturer, price, model number (assumed by manufactured), and an internal item number (assigned by the store).

(*iv*) each manufacturer is represented. The data about a manufacturer are its name, address, items supplied to the store, and their prices.

Give an E-R diagram for the database. Note that some information may be represented by attributes other information many be represented by relationship ?

Solution :

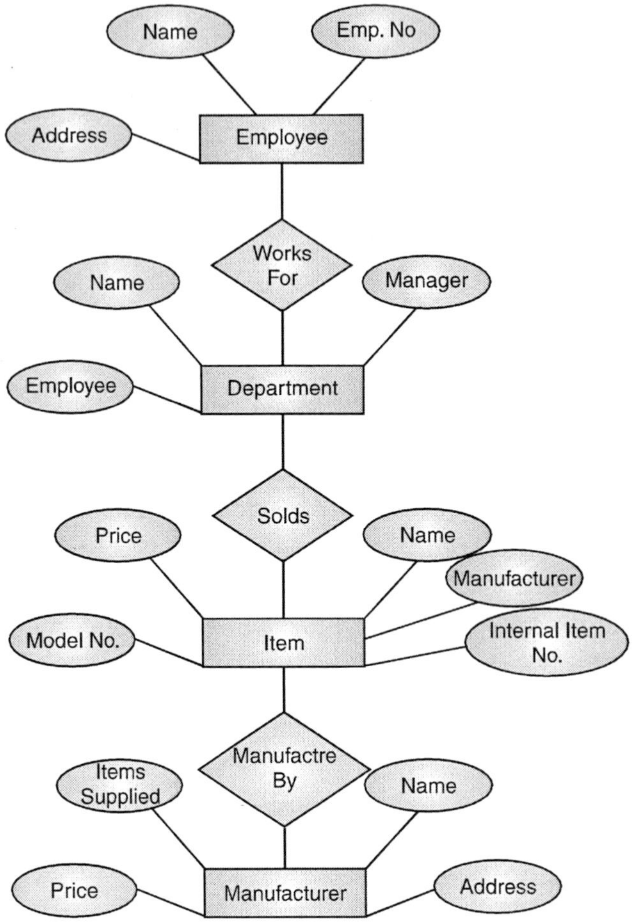

Fig. 2.34. E-R Diagram

Summary

- The Entity-Relationship Model is a high level data Model based on a perception of a real world that consists of a collection of basic objects, called entities and of relationships among these objects.
- An entity can be define as a 'thing' or 'object' in the real world that is distinguishable from all other objects, whereas an entity set is a set of entities of the same type that share the same properties or attributes.
- Attributes are the descriptive properties passed by each member of an entity set.
- The domain of an attribute is the collection of all possible values, and attribute can have.
- Mapping constraints express the number of entities to which another entity can be associated via a relationship set.

- A key is a value which can always be used to uniquely identify an object instance. It is of five types : super key, candidate key, primary key, composite and foreign key.
- Generalization is the abstracting process of viewing set of objects as a single general class by concentrating on characteristics of lower level entities while ignoring there is different.
- Aggregation is a technique to express relationship among relationship.
- Specialization is the process of designing subgroupings with an entity set.
- Data Administrator is that particular person who has the central responsibility for the data whereas Database Administrator is a technical person responsible for implementation of data.

Selected Exercises

1. Discuss briefly the convention for displaying an E-R diagram.
2. What is the difference between specialization and Generalization with respect to database?
3. Explain difference between a weak and a strong entity set.
4. Explain the destination among the terms, primary key, candidate key and super key.
5. What is the role of Database Administrator?
6. Construct an E-R diagram for a hospital with a set of patients and a set of medical doctors. Associate with each patient, a log of various test and examination conducted.
7. Explain the mapping constraints in detail.
8. Discuss the concept of Aggregation. Give two examples where this concept is useful ?
9. Discuss the Relationship and its types ?

❒❒❒

Chapter

3 Introduction to Data Model

Inside this Chapter

3.0. INTRODUCTION OF DATA MODEL

To understand th *RBDMS* first of all we have to have an idea of about what is a data model for what purpose a data model is used ? A data model is basically a collection of concepts that are used to specify the overall structure of the database. When we say structure of the database we means that the data types, relationships that exists and the constraints that should be applied on the data to be stored under that data model basically data models an categorised into 3 categories :

1. Physical data model/low level data model
2. High level/object based logical data model
3. Record based logical model

Relational data model is an example of record based logical model in which data is described both at logical and view level. This record based logical data model is used to specify the overall logical structure of the data based.

E.F. codd introdued the relational model in 1970, the relational model has 3 well defined components that are as follows :

1. the relational table (logical data structure) which stores actual data.
2. to maintain the consistency of the databased, some set of integrity rules.
3. set of operations that allows how the data are manipulated.

3.1. RELATIONAL MODEL CONCEPTS

(*i*) A table (synonym for relation proposed by *E.F.* codd) is a logical construct in *RDBMS*. It is a 2 *D* structure having rows and columns. Rows of a table represents the entities of an entity set in *ER* model and column represents the attributes in *ER* model. *i.e.* each row/column intersection gives a single data value.

(*ii*) In *RDBMS* rows are called as tuples.

(*iii*) Values within a single column are homogeneous *i.e.* all tuples are having same data format for a single column.

e.g. if the data type of a column is integer then all the tuples in that relation will have integer type values for that column.

(*iv*) All column have a specific range of values this is called the attribute domain.

(*v*) Each tuple in a relation is unique.

(*vi*) The columns of a table/relation are assigned a distinct name.

(*vii*) The ordering the tuples in a relation is immaterial.

(*viii*) The sequence of columns within a relation is of no importance.

(*ix*) Each table must have a primary key. Primary key is a principle mean in a relation in order to identify tuples in a relation.

e.g.

STUDENT Relation Columns/attribute

M/01	Abhinav	73, Ghaziabad	28	28 Jan. 77
M/02	Ajay	86, Saket	31	15 Mar. 77
M/03	Asan	71, Pushp Vihar	33	16 Apr. 75
M/04	Rabi	15 Rajouri	33	20 May 75

tuples

The **schema** of the above relation is

STUDENT (**ENROLL#**, Name, Add, Age, dob)

Relation name relation atributes

In general, a relation schema R is defined as $R(A_1, A_2, A_3, ..., A_n)$

where $A_1, A_2, A_3, ..., A_n$ are the attributes and the relation R is called a relation of degree n.

The values at any time stored in a table is called as the relation data base state/ instance of the data base. In the above example instance of the STUDENT has 4 tuples.

Note : Instance/database state changes every time you manipulate the database.

3.2. KEYS

Basically key is a set of one or more attributes that uniquely identifies tuples in a relation. If key is having only one attribute then it is a simple key, if it is having more than one attribute then such keys are called as **composite** keys.

An attribute that is part of key is known as **key attribute.**

In the previous example Enroll# of each student is different so Enroll# uniquely identify the tuples we can say that

$$\text{Enroll\#} \longrightarrow \text{Name, Add, Age, dob}$$

i.e. if we know the values of Enroll# we are able to look up the values of Name, age, add, dob of that student.

⇒ Every relation hast at least one default key set of all its attributes this is called default **super** key.

$$\text{SUPER KEY } (R) = \{A_1, A_2 \text{ ...... } A_n\}$$

⇒ A candidate key is a minimal super key.

e.g. [Enroll#, name] is a super key but not a candidate key because Enroll# by itself is capable of identifying tuples uniquely, thus the candidate key is enroll#.

Similarly {Name, add} is also a candidate key because No 2 students having same name reside on same address.

Thus, a relation can have **more than one** candidate key.

Out of these candidate keys of a relation *R*, the designer of the database chooses one key as a principle means to identify the tuples in relation *R* and that is known as **primary key** and remaining are called as alternate key.

It means primary key (choosen by the designer) is a super key as well as candidate key but reverse is not true.

3.2.1. Secondary Key

Secondary key is defined as the key that is used strictly for data retrival purposes. *e.g.* most of the student does not remember their enrollment number. So if a student tills his/her names and dob then this {Name, dob} is a secondary key. Secondary key may match more than one tuple but secondary key narrows down the search.

3.3. CODD'S RELATIONAL DATABASE RULES

Dr E.F. Codd published a list of 12 rules in 1985 in order to define a relational database system. The present dominant database vendors do not fully support all 12 rules. Basically, a DBMS can be regarded as *RDBMS* if it obeys 3 rules.

→ information rule

→ Guaranteed access rule

→ all relationships between data must be represented explicitly in that data itself.

Rule 1 : Information rule says that all the information in a *RDBMS* must be logically represented as a column values in a row within tables.

Rule 2 : Guaranteed Access says that all data should be accesible without ambiguity. It should be accessible through a combination of table name, primary key value and column name.

Rule 3 : Systematic treatment of Nulls says that Null values must be represented and treated in a standard way. It must be different from zero or spaces and should be independent of data type.

Rule 4 : Dynamic online catalog based on Relational Model says that *RDBMS* maintains metadata as ordinary data *i.e.*, in the tables within the database. These tables are created, owned and maintained by the *DBMS* itself.

Rule 5 : Comprehensive data subrange rule says that database must support at least one well defined declartive language that is used for data definition, view defined, data manipulation, integrity constraints and transaction management.

Rule 6 : View updating rule says that any view that is theoretically updatable must be updatable through the system. Each view should support the same full range of data manipulation.

Rule 7 : High level insert, update and delete states that inert, update and delete operations should be supported for any retrievable set rather than just for a single row in a single table.

Rule 8 : Physical data independance states that whenever the physical access methods and/or storage structures are changed, the application programs and all ad-hoc facilities remain unchanged. Changes to physical schema can be made without affecting how the user access it.

Rule 9 : Logical data independence states that whenever the logical schema (table structure) changes, the application programs and ad-hoc facilities remains unchanged.

Rule 10 : Integrity independence says that the relational database language must be capable of defining all integrity constraints and they are stored in the system catalog or data dictionary.

Rule 11 : Distribution independence states that the application programs as well as the end users are unware of the location of the data (*i.e.* whether data is stored locally or distributed).

Rule 12 : Non subversion rule zero says that if system support any low level access to the database it should be through integrity rules.

3.4. RELATIONAL MODEL CONSTRAINTS OR INTEGRITY CONSTRAINTS

Integrity means accuracy or correctness of the data. To maintain the consistency of the database and to insure that changes to the database are made through authorised users, there are integrity constraints/conditions that every DBA should define. The basic purpose of these constraints is to ensure that the consistency of the database is maintained after a change is made to it.

3.4.1. Entity Integrity

Entity integrity states that all primary key enteries are unique and no entery for primary key can be null this is because the value of primary key is used to identify individual tuple in a relation; and if null values are allowed for primary key we will not be able to identify tuples having null values for primary key.

e.g. if 2 tuples in STUDENT relation is having null value for column Enroll#, then will not be identify/differentiate between them.

3.4.2. Domain Constraint

Domain constraints specifies that the domain of each attribute A must be atomic in nature and attribute A must take only a single value from that atomic domain.

i.e. dom (*A*) is a set of permissible values which must be atomic in nature.

We have standard domain types such as integer type, character type, data etc.

3.4.3. Referential Integrity

Referential integrity is specified on 2 relations.

Referential integrity is due to the presence of **foreign key.**

Let us consider an example having 2 relation

STUDENT

Primary Key → Enrol#; Course# → Foreign key

Enrol#	*Name*	*Course#*
101	Ram	10
102	Shyam	14
103	Hari	12

COURSE

Primary Key

Course#	*Name*	*Duration*
10	B. Tech	4
14	MCA	3
15	MBA	2

Student relation shows all the students that are enrolled for various course where as course relation gives the detailed information about each course. There are 2 dangling tuples situations

(*a*) in student relation, 3rd tuple (Enroll# = 103) is having 12 as Course#, corresponding to which there is no entry in course relation.

(*b*) in course relation, 3rd tuple (Course# = 15) is not having any entry in STUDENT relation.

Obviously, **first kind** of situation is not desirable because unless untill a course is there No student can be enrolled into it.

Here Course in COURSE relation is the **primary key** of the relation whereas Course# in STUDENT is the **foreign key** of the STUDENT relation.

Let $r_1(R_1)$ and $r_2(R_2)$ be the 2 relations with primary keys PK_1 and PK_2 respectively. We say that a subset α of R_2 is a foreign key with reference to PK_1 in $r_1(R_1)$ if it is necessary that for every t_2 in $r_2(R_2)$ there must be a tuple t_1 in $r_1(R_1)$ *s.t.* $t_1[PK_1] = t_2[\alpha]$.

3.4.4. Not Null Constraint

States that every row in the table has a value for that column is specified on column 4.

3.4.5. Unique Constraint

Unique constraint states that no duplicate values exist for that column.

3.4.6. Key Constraints

Key constraints states that for a relation $R(A_1, A_2, \ldots, A_n)$, no 2 tuples can have the same combination of values for all their attributes.

3.5. RELATIONAL ALGEBRA

Relational algebra defines the theoretical way of manipulating tables contents. It provides basic set of operations for relational data model.

One or more relations ⟶ [Relational Algebra] ⟶ New relation

Consists of set of operations

Operation are categorised as

(*A*) *Fundamental operations* includes select, project, union, set difference, cartesian product, rename.

(*B*) Other operations that includes set intersaction, natural join, division, assignment.

3.5.1. Select

Select is a unary operator because it operates on one relation and is represented by Greek letter sigma (σ). Sigma selects tuples that satisfies a given predicate. General form,

argument relation

σ (*R*)

Selection condition

↑

Predicate

Example, find all students whose age is greater than 25

σ (STUDENT)

age > 25

Operations for predicates are =, ≠, <, <=, >, >= whereas connectives are ∧ (AND), ∨ (OR), > (logical NOT) *e.g.* find all students who are enrolled in course no. 14 and having age greater than 25.

σ (STUDENT)

age > 25 ∧ Course# = 14

Degree of the argument relation *R* = degree of resulting relation, i.e. the resulting relation will have same number of attributes as argument relation.

3.5.2. Project

Project is a unary operator that filterout certain attributes from argument relation. It is represented by greek letter Pi (π). Duplicate tuples are eliminated from the resulting relation.

General form

π (*R*)

<attribute list>

Example : Find enrolment and name of all the students.

Solution : σ (STUDENT)

enroll#, Name

Note : Select and project can be combined together.

Example : Find enroll# and name of all the students who is above 22.

$$\pi_{\text{enroll\#, Name}} (\sigma_{\text{age > = 22}} (\text{STUDENT})]$$

STUDENT

Enroll#	*Name*	*add*	*age*	*dob*	*course#*
101	RAM	Delhi	32	12 Dec. 77	14
102	AMIT	Delhi	20	12 Jan. 89	17
103	ARUN	Kolkatta	23	11 Feb 86	15

Original relation

⇓ after σ age > = 22

Enroll#	*Name*	*add*	*age*	*dob*	*course#*
101	RAM	Delhi	32	11 Dec. 77	14
103	ARUN	Kolkata	20	11 Feb. 86	15

⇓ after π enroll#, Name

Enroll#	Name
101	RAM
103	ARUN

3.5.3. Union Operation

If there are 2 relations $R(A_1, A_2, ..., A_n)$ and $S(B_1, B_2, ..., B_n)$ then they are said to be union compatible if they have same degree n and if the dom (A_i) = dom (B_i) for $1 \leq i \leq n$.

Therefore, for domain compatible relations R and S, the UNION of R, S, denoted by $R \cup S$; is a relation that includes all tuples that are either in R or in S or in both.

Example : find name of all students along with teacher name.

Solution : π_{Name} (STUDENTS) $\cup$ π_{Name} (TEACHER)

3.5.4. Intersection

The result of this operation, denoted by $R \cap S$, is a relation that includes all tuples that are both in R and S. It is represented by $\cap$ and is a binary operation.

Suppose there are schemas BORROWER (Loanno, custname, branch) and Depositor (Acno, cust name, bal).

Example : Find name of all customers who have a loan and account.

$\pi_{\text{Cust name}}$ (BORROWER) $\cap$ $\pi_{\text{Cust name}}$ (DEPOSITOR)

3.5.5. Set Difference

The operation, denoted by —, allows us to find tuples that are in one relation but not in another relation.

3.5.6. Cartesian Product

Cartesian product, denoted by X, allows us to combine information from any 2 relations.

Let $r_1(R_1)$ and $r_2(R_2)$ be the relations. Their $r_1 \times r_2$ is the cartesian product of r_1 & r_2.

If R_1 $(A_1, A_2, A_3, ..., A_n)$ and R_2 $(B_1, B_2, ..., B_m)$.

Then $R_1 \times R_2$ will have $n + m$ attributes and $nR_1 \times nR_2$ tuples where

$$nR_1 = \text{no of tuples in } r_1$$
$$nR_2 = \text{no of tuples in } r_2$$

e.g. STUDENT (Enroll#, Name, Course#)

COURSE (Course#, Course name, Duration)

Enroll#	Name	Course#
101	Amit	10
102	Ramesh	12
103	Ram	10

Course#	Course Name	Duration
10	B-Tech	4
12	MBA	2
14	MCA	3

Here $n = 3, \quad m = 3$

$nR_1 = 3, \quad nR_2 = 3$

so (Student × Course) will have $n + m$, *i.e.*, 3 +3 = 6 attributes & 3 × 3 = 9 tuples.

Then,

STUDENT × COURSE will have 9 tuples and 6 columns

Enroll#	*Name*	*Course#*	*Course*	*Course Name*	*Duration*
101	Amit	10	10	B-Tech	4
101	Amit	10	12	MBA	2
101	Amit	10	14	MCA	3
102	Ramesh	12	10	B-Tech	4
102	Ramesh	12	12	MBA	2
102	Ramesh	12	14	MCA	3
103	Ram	10	10	B-Tech	4
103	Ram	10	12	MBA	2
103	Ram	10	14	MCA	3

3.5.7. Join

Join operation, denoted by ⋈, is a binary operation used to combine information from 2 given relations, it allows us to combine certain selection and a cartesian product into one single operation.

Join man be theta join, Equi join, Natural join.

Theta join is the most general join

$$R \bowtie S = \sigma\,(R \times X)$$

where θ may be =, <, <=, >, >=, ≠

If θ is only =, then join is called **Equi join** and represented as ⋈

Theta/Equi join is having duplicate attributes natural join is an equijoin that projects away the duplicate attributes.

For the previous example of cartesian product

STUDENT ⋈ COURSE

Enrol#	*Name*	*Course#*	*Course Name*	*Duration*
101	Amit	10	B-Tech	4
102	Ramesh	12	MBA	2
103	Ram	10	B-Tech	4

Natural join has 3 more variations.

(*a*) Left outer joint has all the tuples of natural join along with the tuples from left relations for which there is no matching tuples in right relation. It is represent by $R \bowtie S$.

(*b*) Right outer join has all the tuples of natural join along with the tuples from right relations for which there is no matching tuples in left relation, represented by $R \bowtie S$.

(*c*) Full outer join, represented by ⋈, is combination of both left and right outer joins.

Condition for join is that the joining attributes of both relation must have same domain

$$R \underset{A = B}{\bowtie} S$$

Then $\quad \text{dom}(A) = \text{dom}(B)$

3.5.8. Division

Division operation, denoted by ÷ is, used to query that include the phrase **for all.** It requires the use of one single column table and one 2 column table.

relation 1

Course#	*Name*
10	B-Tech
12	MBA
14	MCA

relation2

Course#
10

Then relation ÷ relation 2 ⇒

Name
B Tech

3.5.9. Extended Relational Algebra

(*i*) Aggregate function take a collection of values and returns a single value as a result. It is represented by calligraphic ζ.

e.g. Count the total no of enrolled student

(STUDENT)

G Count (enroll#)

e.g. Count the total no of enrolled students in each course

(STUDENT)

COURSE G Count (enroll#)

First grouping is done on the basic of course# then aggregate function is applied.

Summary

In this chapter we have studied the most important model of dbms – Relational model.

Relational model is an example of record based logical model.

- In *RDBMS*, tables are basic building blocks. Tables consists of rows and columns. Row and columns of *RDBMS* are quivalent to entities and attributes of *ER* model.
- Keys are important in *RDBMS* several types of keys are these primary key, candidate key, foreign key and secondary key.
- Primary key is closed by the database designer as a principle means to distinguish tuples in a table.
- Two tables are related with foreign key.

- E.F. codd formulated a list of 12 rules based upon these rules we can deciders a DBMS to be a RDBMS or not.
- In RDBMS, tables are basic building blocks. Tables consists of rows and columns. Rows and columns of RDBMS are equivalent of entities and attributes of ER model.
- Keys are important in RDBMS. Several types of keys are there i.e., primary key. Candidate key, foreign key etc.
- Primary key is chosen by the database designer as a principle mean to ditinguish tuples in a table.
- 2 tables are relateds with a foreign key.
- RDBMS supports relation algebra functions such as SELECT, PROJECT, UNION, JOIN etc.

Multiple Choice Questions

1. Entities are analogous to
(*a*) Relations (*b*) DBMS
(*c*) Attributes (*d*) Tuples

2. Attributes of ER correspond to
(*a*) Rows of a table (*b*) Degree of a table
(*c*) Column of a table (*d*) None

3. Number of tuples of a relation is called as
(*a*) Degree (*b*) Cardinality
(*c*) Extension (*d*) None

4. A relation with degree, 2 is known as
(*a*) Binary (*b*) 2-ary relation
(*c*) N-ary relation (*d*) None

5. A virtual relation is also known as
(*a*) view (*b*) table
(*c*) snapshot (*d*) none of the above

6. What is the cardinality of table with 100 rows and 10 columns?
(*a*) 100 (*b*) 10
(*c*) 250 (*d*) none

7. Degree of a table with 100 rows and 10 column is
(*a*) 100 (*b*) 10
(*c*) 250 (*d*) none

8. How many tables can be joined to create a view
(*a*) 1 (*b*) 2
(*c*) depends on DBMS (*d*) none

9. A set of tuples at any given time is called as
(*a*) table (*b*) relation
(*c*) extension (*d*) none

10. A key that has no meaning to the business or organization is
(*a*) Candidate key (*b*) Alternate key
(*c*) Artificial key (*d*) None

11. Foreign keys are placed in

(*a*) the parent table
(*b*) the child table
(*c*) either the parent or the chidl table
(*d*) neither the parent nor the child

12. For every relationship, there are How many possible referential integrity actions ?

(*a*) one
(*b*) two
(*c*) six
(*d*) indefinite

13. Which of the following is not true of a null value

(*a*) the value is unknown
(*b*) the value is known to be acceptable
(*c*) the value is not appriciate
(*d*) the value is known to be blank

14. Which of the following attributes are required in a table design ?

(*a*) foreign key
(*b*) primary key
(*c*) All attributes
(*d*) No attribute

15. A primary key should be defined with :

(*a*) NULL
(*b*) NOT NULL
(*c*) Both (*a*) and (*b*)
(*d*) neighter (*a*) nor (*b*)

Selected Exercises

1. What is the difference between a relation and a table ?

2. Why are integrity constraints important in a database ?

3. Differentiate between natural join, Theta join and Equi join.

4. Define *RDBMS* and its properties ?

5. (*a*) Explain the differences between a primary key candidate key and super key

(*b*) What are the restrictions that are put on update operation using a view?

6. List the reasons why Null values might be introduced into the database.

7. (*a*) What are unions? When 2 tables aer said to be union compatible. How are the results ordered in a union compatible?

(*b*) State and explain entity integrity and referential integrity.

8. (*a*) Describe the differences in meaning between the term relation and relation schemme.

(*b*) What is a view? What are its advantages? Explain the syntax for creating views.

(*c*) What are integrity constraints? Explain each of them.

9. Consider the following relations.

Employee (Emp-name, street, city)

Works (Emp-name, company-name, salary)

Company (company-name, city)

Manages (emp-name, manager-name)

Write the following queries in relational Algebra.

(*a*) Find the names of all employes who live in the same city on the company for which they work.

(*b*) Find the names of all employees who lives in the same city as do their managers.

(*c*) Find the names and cities of residence for all employes who work for company "XYZ".

(*d*) Find the name of city where employee '*X*' works.

10. Given the relation schme

Enroll (S#, C#, section)
Teach (Profid, C# section)
Advise (profid, S#)
Pre-req (C#, Pre-C#)
Student (S#, C#, grade, year)
Student (S#, Sname)
Gives queries expressed in Tuple relational calculus.

(*i*) List of students taking courses with smith and/ones
(*ii*) List of all students taking at least one course that their advise teaches.
(*iii*) List all student number and course number.
(*iv*) List all the student and course number who got grade 'A'.

11. Consider the following relations.
supplier (sid, name, address)
part (pid, pname, color)
catalog (sid, pid, cost)
gives queries in relation algebra

(*a*) find names of supplies who supply red parts.
(*b*) find names of supplies who supply red or green parts
(*c*) find names of supplies who supply red and green parts
(*d*) find name of supplies who supply red but not green parts

12. Differentiate between natural join, theta join and equi join with example.

13. Give some properties of relations?

14. Consider the following database and answer the queries in relational algebra.
Branch (branch no., street, city, pincode)
Staff (staff no., name, position, sex, dob, sal, branch number)
property for rent (property no, street, city, pincode, type, roms, rent, owners no, staff no, branch no)
client (client no., name, phone, pretype, maxrent)
private owner (owner no, name, address, comment)
viewing (client no, property no, vicodata, comment)
registration (client no, branch no, staff no., data joined)

(*a*) List the names and comments of all the clients who have viewed a property.
(*b*) List all cities where there is either a branch office or a property for rent.
(*c*) List all cities where there is a branch office but no properties for rent.
(*d*) identify all clients who have viewed all properties with 3 rooms.
(*e*) produce a status report on property viewing.

15. Consider 2 relations *R* and *S* containing n_1 and n_2 tuples respectively with $n_1 > n_2$ 0. Give the minimum and maximum possible sizes (in tuples) for the result relation produced by each of the following relational algebra expressions.

(*a*) RS
(*b*) R/S
(*c*) R × S
(*d*) S/R

❑❑❑

Chapter

4 Introduction to SQL and its Components

Inside this Chapter

4.1. INTRODUCTION

STRUCTURED QUERY LANGUAGE *(Imp.)*

SQL stands for structured Query language. It is non-procedural in nature. The SQL standard specifies data definition, data manipulation, and other associated facilities of a DBMS that supports the relational data Model. SQL is a comprehensive language for controlling and interacting with a database management system. This language was named as the structured English Query language (SEQUEL) the name later was shortened to structured query language.

SQL can also be define as a query language that allows access to data residing in a database management system.

SQL plays many different roles:

— It is an interactive language because the user gains occurs to the data by issuing desired query. This provides a convenient, easy to use to for ad-hoc database queries.

— It is a database programming language because SQL commands can easily be embedded into application programs to access the data from a database.

— It is a database administration language because the database administrator uses SQL to define the database structure and control access to the stored data.

— It is a client/server language because it can be used to access remote (over a network) database.

— It is a distributed database language because SQL helps in distribution of data across many connected computer system.

Advantages of SQL :

1. **Standard Independent Language:** The universal rules of the SQL have been established by ANSI and ISO. Therefore, the SQL is an open language implying it is not owned or controlled by any single company.
2. **Speed:** The intense competition among database vendors has resulted in faster, more robust database management system that work at lower costs per transaction.
3. **Cross-Platform Abilities:** One of the biggest hardships of using a programming language to access database is that it seldom produces a "true" cross platform application. SQL may be relatively new to some of the most popular programming languages like COBOL, or C, but it has been in use on different hardware platforms for years.
4. **Easy to Learn and use:** SQL Statements resemble simple English sentences, making SQL easy to learn and understand.
5. **Less programming:** SQL allows extraction, manipulation, and organisation of data with less programming as compared to traditional methods.

4.1.1. Tables

In relational database systems (DBS) data are represented using *tables* (*relations*), A query issued against the DBS also results in a table. A table has the following structure:

Column 1	Column 2	...	Column *n*
...	...	...	...

← Tuple (or Record)

A table is uniquely identified by its name and consists of *rows* that contain the stored information, each row containing exactly one *tuple* (or *record*). A table can have one or more columns. A *column* is made up of a column name and a data type, and it describes an attribute of the tuples. The structure of a table, also called *relation schema*, thus is defined by its attributes. The type of information to be stored in a table is defined by the data types of the attributes at table creation time.

SQL uses the terms *table*, *row*, and *column* for *relation*, *tuple*, and *attribute*, respectively. In this tutorial we will use the terms interchangeably.

A table can have up to 254 columns which may have different or same data types and sets of values (domains), respectively. Possible domains are alphanumeric data (strings), number and date formats. Oracle offers the following basic data types:

- **char**(*n*): Fixed-length character data (string), *n* characters long. The maximum size for *n* is 255 bytes (2000 in Oracle8). Note that a string of type **char** is always padded on right with blanks to full length of *n*. (☞ can be memory consuming).
  Example: **char**(40)
- **varchar 2**(*n*): Variable-length cahracter string. The maximum size for *n* is 2000 (4000 in Oracle8). Only the bytes used for a string require storage. *Example*: **varchar2**(80)

- **number**(*o*, *d*): Numeric data type for integers and reals. *o* = overall number of digits, *d* = number of digits to the right of the decimal point.

  Maximum values: *o* = 38, *d* = – 84 to + 127. *Examples*: **number**(8), **number**(5, 2)

  Note that, *e.g.*, **number**(5, 2) cannot contain anything larger than 999.99 without resulting in an error. Data types derived from **number** are **int**[**eger**], **dec**[**imal**], **smallint** and **real**.
- **date**: Date data type for storing data and time.

  The default format for a date is: DD-MMM-YY. *Examples*: '13-OCT-94', 07-JAN-98'.
- **long:** Character data up to a length of 2GB. Only one **long** column is allowed per table.

Note: In ORACLE-SQL there is no data type **boolean**. It can, however, be simulated by using either **char**(l) or **number**(l).

As long as no constraint restricts the possible values of an attribute, it may have the special value null (for unknown). This value is different from the number 0, and it is also different from the empty string ''.

Further properties of tables are:

- the order in which tuples appear in a table is not relevant (unless a query requires an explicit sorting).
- a table has no duplicate tuples (depending on the query, however, duplicate tuples can appear in the query result).

A *database schema* is a set of relation schemas. The extension of a *database schema* at database run-time is called a *database instance* or *database*, for short.

4.2. SQL*PLUS

4.2.1. Introduction

SQL*Plus is the interactive (low-level) user interface to the Oracle database management system. Typically, SQL*Plus is used to issue ad-hoc queries and to view the query result on the screen. Some of the features of SQL*Plus are:

- A built-in command line editor can be used to edit (incorrect) SQL queries. Instead of this line editor any editor installed on the computer can be invoked.
- There are numerous commands to format the output of a query.
- SQL*Plus provides an online-help.
- Query results can be stored in files which then can be printed.

Queries that are frequently issued can be saved to a file and invoked later. Queries can be parameterized such that it is possible to invoke a saved query with a parameter.

A Minimal User Guide

Before you start SQL*Plus make sure that the following UNIX shell variables are properly set (shell variables can be checked using the **env** command, e.g., **env** | **grep** Oracle):

- ORACLE_HOME, e.g., ORACLE_HOME=/usr/pkg/oracle/734
- ORACLE_SID, e.g., ORACLE_SID=prod

In order to invoke SQL*Plus from a UNIX shell, the command **sqlplus** has to be issued. SQL*Plus then displays some information about the product, and prompts you for your user name and password for the Oracle system.

```
gertz(catbert)54: sqlplus

SQL*Plus: Release 3.3.4.0.1 - Production on Sun Dec 20 19:16:52 1998

Copyright (c) Oracle Corporation 1979, 1996. All rights reserved.

Enter user-name: scott

Enter password:

Connected to:

Oracle7 Server Release 7.3.4.0.1 - Production Release

With the distributed option

PL/SQL Release 2.3.4.0.0 - Production

SQL>
```

SQL> is the prompt you get when you are connected to the Oracle database system. In SQL*Plus you can divide a statement into separate lines, each continuing line is indicated by a prompt such 2>, 3> etc. An SQL statement must always be terminated by a semicolon (;). In addition to the SQL statements discussed in the previous section, SQL*Plus provides some special SQL*Plus commands. These commands need not be terminated by a semicolon. Upper and lower case letters are only important for string comparisons. An SQL query can always be interrupted by using <Control>C. To exit SQL*Plus you can either type **exit** or **quit.**

Editor Commands

The most recently issued SQL statement is stored in the SQL buffer, independent of whether the statement has a correct syntax or not. You can edit the buffer using the following commands:

- l[ist] lists all lines in the SQL buffer and sets the current line (marked with an "*") to the last line in the buffer.
- l<number> sets the actual line to <number>
- c[Change]/<old_string>/<new_string> replaces the first occurrence of <old_string> by <new_string> (for the actual line)
- a[ppend] <string> appends <string> to the current line
- del deletes the current line
- r [un] executes the current buffer contents
- get<file> reads the data from the file <file> into the buffer
- save<file> writes the current buffer into the file <file>
- edit invokes an editor and loads the current buffer into the editor. After exiting the editor the modified SQL statement is stored in the buffer and can be executed (command r).

The editor can be defined in the SQL*Plus shell by typing the command **define _editor** = <name>, where <name> can be any editor such as **emacs, vi, joe,** or **jove**.

SQL* Plus Help System and Other Useful Commands

- To get the online help in SQL*Plus just type **help** <command>, or just **help** to get information about how to use the **help** command. In oracle Version 7 one can get the complete list of possible commands by typing **help command.**
- To change the password, in oracle Version 7 the command

  alter user <user> **identified by** <new_password>;

  is used. In oracle Version 8 the command **passw** <user> prompts the user for the old/new password.
- The command **desc[ribe]** <table> lists all columns of the given table together with their data types and information about whether null values are allowed or not.
- You can invoke a UNIX command from the SQL*Plus shell by using **host** <UNIX_command>. For example, **host Is -la *.sql** lists all SQL files in the current directory.
- You can log your SQL*Plus session and thus queries and query results by using the command **spool** <file>. All information displayed on screen is then stored in <file> which automatically gets the extension . 1st. The command **spool off** turns spooling off.
- The command **copy** can be used to copy a complete table. For example, the command

  copy from scott/tiger **create** EMPL **using select * from** EMP;

  copies the table EMP of the user scott with password tiger into the relation EMPL. The relation EMP is automatically created and its structure is derived based on the attributes listed in the **select** clause.
- SQL commands saved in a file <name>.sql can be loaded into SQL*Plus and executed using the command @<name>.
- Comments are introduced by the clause **rem[ark]** (only allowed between SQL statements), or - - (allowed within SQL statements).

Formatting the Output

SQL*Plus provides numerous commands to format query results and to build simple reports. For this, format variables are set and these settings are only valid during the SQL*Plus session. They get lost after terminating SQL*Plus. It is, however, possible to save settings in a file named login.sql in your home directory. Each time you invoke SQL*Plus this file is automatically loaded.

The command **column** <column name> <option 1> <option 2> ... is used to format columns of your query result. The most frequently used options are:

- **format** A<n> For alphanumeric data, this option sets the length of <column name> to <n>. For columns having the data type **number,** the **format** command can be used to specify the format before and after the decimal point. For example, **format** 99,999.99 specifies that if a value has more than three digits in front of the decimal point, digits are separated by a colon, and only two digits are displayed after the decimal point.

- The option **heading** <text> relabels <column name> and gives it a new heading.
- **null** <text> is used to specify the output of null values (typically, null values are not displayed).
- **column** <column name> **clear** deletes the format definitions for <column name>.

The command **set linesize** <number> can be used to set the maximum length of a single line that can be displayed on screen, **set pagesize** <number> sets the total number of lines SQL*Plus displays before printing the column names and headings, respectively, of the selected rows.

Several other formatting features can be enabled by setting SQL*Plus variables. The command **show all** displays all variables and their current values. To set a variable, type **set** <variable> <value>. For example, **set timing on** causes SQL*Plus to display timing statistics for each SQL command that is executed, **set pause on** [<text>] makes SQL*Plus wait for you to press **Return** after the number of lines defined by **set pagesize** has been displayed. <text> is the message SQL*Plus will display at the bottom of the screen as it waits for you to hit **Return**.

4.3. INTRODUCTION TO SQL

SQL is an acronym for structured Query language.

* Provides an interface to RDBMS.
* Non procedural in Nature
* Has 4 Components namely DML, DDL, DCL, DQL

(*a*) DML → data manipulation language used to change the data within the data.
Commands are as follows

(*i*) Insert - to insert new Tuple(s) in the Table.
(*ii*) Update - to modify the Tuple(s) in the Table.
(*iii*) Delete - to delete Tuple(s) from the table,

(*b*) DDL → data definition language used to create, modify & delete the database structures not the data.
Commands are as follows

Create - to create objects like table, index, view
Drop - to drop the entire object
Alter - to modify the structure of the table, constraints.

(*c*) DCL → data control language controls access to data & to the database.
Commands as follows

(*i*) Commit - to make changes persistent to the database
(*ii*) Rollback - to severe back the changes
(*iii*) Gront → granting permissions to users
(*iv*) Revoke → reverting the permissions back

(*d*) DQL → data query language used to retrieve data from the database.
Command is —

Select - retrieving the tuple(s) from the table

4.3.1. DQL (Data Query Language)

To select all rows & all column of a table.

Select <Col 1> to <COI N>

From <Tablename>;

or

Select * **(* represents all columns)**

From <Tablename>;

4.3.2. Selected columns & all rows

Select Col 5, Col 9

From < Tablename>;

4.3.3. Selected rows & all columns, have to use where clause

Select *

From (Employee)

Where age = 10;

Select *

from <Tablename>

Where <Conditions>;

4.3.4. Selected col^n & selected rows

Select <col 2>, <col 7>

From <Tablename>

Where <conditions>;

4.3.5. Eliminating duplicate rows

Table may hold duplicate rows. DISTINCT clause allows removing duplicat from the result set duplicate rows are eliminated from the resultset ot from the actual table.

Select DISTINCT <col name>

From <Tablename>;

4.3.6. Sorting Data

Data can be viewed in a sorted order.

Select *

Example Select enfrom employee order by SQL ;

from <Tablename>

Order by <Col 1>, <Col 2> <[Sort order].;

Sort order ASC → default

DESC

4.3.7. Group by having clause

Group by clause groups rows based on distinct values for specified columns.

Select *

from <Tablename>

Group by < Col 1>, <Col 2>, <Col 3>;

Example 4.1. **How many employees are there in each dept.**

Solution.

Select Count (Emp no) ← Select count (emp no)
from emp from emp;
Group by deptro; ↓

First groups based on deptro is toemed then count is preformed.

Example 4.2. **Tells total no. of employees.**

Solution. Having is used in conjuction with group by **clause.** Having clause imposes a condition on the Group by clause.

Example 4.3. **Find minimum & max salary of clerks for each department.**

Solution. Select deptno, min(sal), Max(sal) from emp.
Where job-'clerk' Group by deptno Having count (*)>3;

Complete syntax :

Select [DISTINCT] <column 1>, <column 2> ... <column n>
from <table>
[Where <condition>]
[Group by <col 1>, <col 2>
[Having <conditions>]
[Order by <col 1>, <col 2> ...[asc/DESC]>]
Where [ ] indicates that the clause is optional.

4.4. ORACLE TABLE – DUAL

Oracle provide a workable Table, having only one row and one column, called dual, dual is owned by sys. Dual can be used as in many ways.

(A) Select 2*2 from dual;

2 * 2
4

(B) Select SYSDATE from dual;

SYDATE
04-JUL-2009

(C) Select UID from dual;

UID
60

gives on integer value corresponding to the userid of the currently logging in user.

Aggregate functions

They are statistical function & used to compute single value from of set of attributes values on a column.

1. Count (<attribute>) → counts the number of rows in the table
2. Max (<attribute>) → maximum value of a column
3. Min (<attribute>) → minimum value of column.

4. Sum (<attribute>) → compute sum of values for that column.
5. Avg (<attribute>) → compute the average value of that column.

Example 4.4. Find total number of employees.

Solution.

```
select Count (*)
from emp;
```

Example 4.5. Find total number of employees in each department.

Solution.

```
Select count (*)
from emp
Group by deptno;
```

Example 4.6. Find the difference between the maximum salary & minimum salary of employees.

Solution.

```
Select Max (sal) - Min (sal)
from emp;
```

Different operator & functions

Different operators & functions can be applied on select statement they are as follows.

4.4.1. Numbers functions

In this category various mathematical functions are there such as, abs, cos, sin, exp, power, mod, sqrt, FLOOR, CEIL, ROUND etc

e.g., POWER (2, 3) gives 2^3

ROUND (4. 5,0) gives 5

Example Select abs (-5) from dial;

O/P is | 5 |

4.4.2. String functions

UPPER (<string>) - converts lowercase string into uppercase

LOWER (<string) - converts uppercase string into lowercase.

INITCAP (<string) - converts initial letter of the word into capital letter

LENGTH (<string>) - finds the length of the string

REPLACE (<string 1, search string, string 2.),
- find search string in string 1 & replace it with string 2

SUBSTR (<string>, *n*,[*m*])
- clipout *n* characters from the string starting at position *n*. If *m* is not specified, end of the string is assured.

TO-CHAR (<number) - returns a character string from a numeric value

4.4.3. Operators

+, –, /, *, =, >, >=, <=, AND, OR, NOT, IN, !=

4.4.4. Comparison Operators

(*a*) Null value → to check whether column is NULL or Not <column> **IS** [NOT] NULL

Example Select *
From emp
Where deptno is NULL;

(*b*) Set condition → to find value within a set
<column> [COT] IN (<List of values>)

Example 4.7. Find the details of departments of department number 30 & 50.

Solution.

```
Select  *
from    dept
Where   deptno  In (30,50);
```

Example 4.8. Find details of department other than 20, 40.

Solution.

```
Select  *
from    dept
Where   deptno NOT IN (20, 40);
```

(*c*) BETWEEN → it allows us to give range for conditions
<column> [NOT] BETWEEN <lower bound> AND <upper bound>

Example 4.9. Find employee between the age of 25 & 35.

Solution.

```
Select  *
from    emp
Where   age BETWEEN 25 AND 35;
```

(*d*) LIKE → Like operator is used in conjunction with wildcard to find patterns within string attributes. Wild cards are

Percentage % means any and all following characters
underscore - means any one character

Example 4.10. Find employees whose name contains 'AM' somewhere.

Solution. O/P figure 4.1

```
Select  *
from    emp
Where   ename LIKE '% AM%';
```

Example 4.11. Find employes whose ends with AM & is of only 3 characters.

Solution. output figure 4.2

```
Select  *
from    emp
Where   ename LIKE '-AM;
```

O/P for both queries are

Emp no	*Ename*	*Job*	*MGr*	*HDate*	*Sal*	*Dno*
1011	RAM	Clerk	1061	17-Dec-09	1800	20
1061	SHYAM	MANAGER		10-feb-95	20000	20
1060	AMIT	ANALYST		10-Mar-97	18000	30

Fig. 4.1.

Emp no	*Ename*	*Job*	*MGr*	*HDate*	*Sal*	*Dno*
1011	RAM	Clerk	1061	17-Dec-09	1800	20

Fig. 4.2.

4.5. COMMON DATA TYPE

Data type	format	Description
1. Numeric	Number (*n*,*m*)	– Number (7, 2)-may be upto six digits long including the sign & decimal place with 2 decimal place
	INTEGER	– counting numbers
	SMALLINT	– like INTEGER but limited to integer values upto six digits.
	DECIMAL (*n*, *m*)	– Like NUMBER, but the storage length is minimum specification, *i.e.*, greater length are acceptable.
2. Character	CHAR (*n*)	– fixed length character upto 255 characters
	VARCHAR (*n*) or VARCHAR 2(*n*)	– variable length character data.
3. Date	DATE	– stores data in the julian date format ('DD-MON-YYYY)

4.6. DDL (DATA DEFINITION LANGUAGE)

DDL statements are used to create/modify the structure of the database. *i.e.*, DDL statements add the description of the database to the system catalog.

4.6.1. Create Table

```
CREATE TABLE <tablename>
        (<column 1> <datatype> [<column constraint],
        <column 2> <datatype> [{column constraint>],
        <column 3><datatype> [<column constraint>],
        .......
        .......
        .......
        [<table constraint>]);
```

→ Creates an empty table.

→ for each column, a name, data type & optional constraint is specified.

→ column name must be unique.

→ Table may have table level constraint (s)

```
Create table employee ⟶ eno      number (4)       NOT NULL,
                        Name     varchar 2 (3)    NOT NULL,
                        Job      varchar 2 (10)   NOT NULL,
                        Sal      number (7, 2),
                        PRIMARY KEY (eno);
```

Another use of create is to create indexes. Max improve the efficiency of searches :
CREATE [UNIQUE] INDEX <inexname> ON <tablename> (col 1,[col 2]);
Example CREATE INDEX -pob ON Employee (dno);

4.6.2. Views

View is a virtual table based on some select query. View does not contains any data rather data is populated whenever a query is executed on that particular view. The table on which view is based is called base Table(s).

CREATE VIEW <view name> AS SELECT <query>;

Example 4.12. Find total salary of each department using view.

Solution : Create view total_sal as

```
Select Sum (sal)
from emp
Group by deptro ;
```

name of the view = total_sql
base table = emp

We have a view whose name is total sal. Now we can write a query on that view in the same way as on base table.

Select * from total sal;

		Total sal	
total salary of dept no	= 10 →	15000	gives total salaries of employees
	= 20 →	35000	working in each department
	= 30 →	40000	

Fig. 4.3.

4.6.3. Constraints

Table definitions include integrity constraints, 2 types of constraints are there

(A) Column level constraints are associated with a single column

(B) Table level constraints are associated with more than one column or on entire table.

(A) Column level constraints

1. NOT NULL → specifies that column does not accept nulls.
2. UNIQUE → is used to ensure that all values in column are unique
3. DEFAULT → is used to assign a default value to column if no value is supplied when row is added.
4. PRIMARY KEY → specifies that the column is the only column working as the primary key.
5. REFERENCES → specifies that column is a foreign key.
6. CHECK → is used to validate data when a value is entered for a column. It is used for user defined conditions that to be applied on column(s).

e.g., CREATE TABLE Employee (

```
eno      number (3) CONSTRAINT Emp-pk Primary Key,
ename    varchar 2 (10),
Sal      number (7, 2);
```

(B) Table Level Constraints :

1. PRIMARY KEY
2. UNIQUE
3. FOREIGN KEY
4. CHECK

Meaning of all is same except they are applied on more than one column not on single.

```
CREATE TABLE Employee (
        eno      number (4)  PRIMARY KEY,
        ename    varchar 2(15)  NOT NULL,
        dob      DATE,
        Sal      number (7, 2),
        dno      number (4) REFERENCES DEPARTMENT (dno),
        CONSTRAINT emp-sal CHECK (sal > 0));
```

The above table employee is having the attributes

(*i*) eno which serve as the primary key

(*ii*) ename which cannot have null values

(*iii*) dno is the foreign key as it is referencing the primary key of department table all the above are column level constraints

(*iv*) the condition emp-sal tells that the attribute sal can not have value less than or equal to zero.

4.6.4. Drop

Drop command removes the object permanently *i.e.*, the structure is deleted.

DROP <object> <tablename> [cascade constraints];

e.g.,

```
DROP    TABLE    Employee;
DROP    INDEX    Emp_sal;
DROP    VIEW     Emp_sal;
```

4.6.5. Modifying Table-Column Definitions

It is possible to modify the structure of a table even if the rows have already been populated with the data into the table. This can be done with ALTER command. You can do the following with ALTER command.

(*i*) Can add a column to a table

(*ii*) Can drop a column from the table

(*iii*) Can add a constraint to the table

(*iv*) Can modify the structure of any column

(*v*) Create & destroy indexes.

(*i*) ALTER TABLE <table name>

ADD(<column name><datatype>
[DEFAULT <value>]
[column Constraint]);

ALTER TABLE employee ADD (mgrssn number (4)); will add mgrssn column to table employee.

(*ii*) ALTER TABLE <tablename> DROP COLUMN <columnname>; employee DROP column Mgrssn; will delete mgrssn attribute from the table.

(*iii*) **to modify the structure of a table** ALTER TABLE <tablename>

MODIFY (<column name> [<datatype>]
[DEFAULT <value>]
[<column constraint>];

(*iv*) to Rename a table you cannot use ALTER TABLE rather you have to use RENAME

RENAME <source table> To <destination name>;

4.7. DML DATA MANIPULATION LANGUAGE

After table has been crated, tuple can be inserted into the table, using INSERT command.

4.7.1. INSERT

3 way insert command can be used

(*a*) INSERT INTO <tablename>[<column 1>, <column n>]
VALUES (<value 1, value 2 .... value n>);

e.g.,

INSERT INTO Employee (eno, name, dob, Hdate, dno)
VALUES (101, 'SMARTH', '10-OCT-1981', '1-Jan-2004',10);

A new tuple is inserted into employee table.

(*b*) We can write a query whose result is inserted as tuple(s) in the table.

INSERT INTO <tablename>[<column>, ...column n>] <query>;

e.g.,

INSERT INTO PERMEMP (e#, Hiredate)
SELECT eno, Hdate
from employee
Where Hdate <'31-DEC-2001;

(*c*) INSERT INTO <tablename> values ('&1','&2','&3','&4');

When the above command is executed, you need not to remember the column name, At run time system will ask user to enter the values for respective columns interactively.

4.7.2. Update Command

It is used to modify the data in the table,

UPDATE <table> SET <columname> = <expression> [,<column name> = <expression>] [Where <condition>];

If where clause is not there then all the rows are updated otherwise only the selected rows (based on the condition in where clause) are updated.

Example 4.13. Update the salaries of all employees by 10%.

Solution.

UPDATE Employee
SET sal = sal*1.10;

Example 4.14. Transfer Ramesh from depno 20 to depno = 30 as Analyst & increase his salary by 5000.

Solution.

UPDATE Employee
SET Job = 'ANALYST', DEPT NO = 30, SAL = Sal + 5000
Where ename = 'RAMESH';

4.7.3. Delete Statement

All or selected rows can be deleted using delete command.

DELETE from <tablename>
[WHERE <condition>];

4.8. SET OPERATORS

Sometimes it is useful to combine query results from 2 or more queries into a single query result.

This can be done by using UNION, INTERSECT, MINUS set operators.

<query 1> <set operator> <query 2>

where set operator may be

(*i*) UNION [ALL] returns all rows that either are in the result of query 1 or in the result of query 2 or both.

(*ii*) INTERSECT gives all rows that appeare in both the results of query 1 and query 2.

(*iii*) MINUS gives the rows that appeares in the result of query 1 but not in the result of query 2.

Example 4.15. Assume that there are 2 tables employee 1 & employee 2 in the database having the following values,

eno	*Name*	*Sal*	*dno*
1	Ram	1000	10
2.	Shyam	2000	10
3.	Ramesh	3000	20
4.	Rahul	4000	30

eno	*Name*	*Sal*	*dno*
1	Ram	1000	10
5	Ritu	2000	20
7	Richa	3000	30

Fig. 4.4.

(*i*) Select *
from Employees_1
UNION
Select * O/P

from Employees_2;

eno	Name	Sal	dno
1	Ram	1000	10
2	Shyam	2000	10
3	Ramesh	3000	20
4	Rahul	4000	30
5	Ritu	2000	20
7	Richa	3000	30

Fig. 4.5. Duplicate row(s) are eliminated.

(*ii*)
```
Select    *
from      Employees_
INTERSECT
Select    *
from      employees_2;
```

Output

eno	Name	Sal	dno
1	Ram	1000	10

Fig. 4.6.

(*iii*)
```
Select    *
from      Employees_1
MINUS
Select    *
from      employees_2;
```

Output

eno	Name	Sal	dno
2	Shyam	2000	10
3	Ramesh	3000	20
4	Rahul	4000	30

Fig. 4.7.

4.9. DCL (DATA CONTROL LANGUAGE)

Modifications of tuples are temporarily stored in the database. They become permanent only after the statement

Commit; is executed.

As long as the user has not issued the commit statement, it is possible to undo modifications since the last commit to do so, execute

Rollback;

Granting and revoking permissions : Grant command is used to assign privileges on objects to various users.

```
Grant    <privileges>
ON       <object name>
TO       <uses name>
         [WITH GRANT OPTION];
```

privileges are

Example Grant select ON EMP To Rajiv;

The above command is issued by the user Amit for granting select permissions on Emp table to the user Rajiv.

ALTER
DELETE
INSERT
SELECT
UPDATE
INDEX

With grant option allows the grantee to in turn grant privileges to other users.

```
REVOKE   <privileges>      Revoke command is used in
ON       <object name>     revert back the privileges.
from     <unsername>
```

Join :

Till now we were able to extract inforamtion only from one single relation. We can extract information stored in different tables with join.

Join can be classified as **inner join** & **outerjoin.** The inner join is the join in which only rows that fulfil a given condition are selected. Join condition may be equality condition(natural join or equi join) or an inequality condition (theta join).

Whereas in outerjoin not only the matching but unmatched attribute values are returned. Outer join may be left, right or full outer join.

(*i*)
```
Select   *
from     E1, E2;
```

It gives **cross product** of E_1 & E_2

or

```
Select   *
from     Employee CROSS JOIN Department;
```

(*ii*) Natural join

Determine the common attributes by finding attributes with same name & compatible data type. Then selects only the rows with common values for the common attributes & finally removes the duplicates rows.

```
      SELECT   *
      From     Employee NATURAL JOIN Department;
or    Select   *
      from     Employee Join department using (dno);
```

This query returns only the rows with matching values in the column indicated in the using clause.

```
Select   *
from     employee join department
ON       employee dno = department, d_number;
```

When joining tables do not have commen attribute names then we have to use **JOIN ON** clause.

Employee

eno	*Name*	*Sal*	*dno*	*Mgro*
101	Ram	1000	10	
102	Shyam	2000	10	101
103	Rahul	3000	20	101

Department

d-number	*dname*	*dloc*
10	Research	Delhi
20	Sales	Kolkata
30	Marketing	Delhi

Employee Join Department

eno	*Name*	*Sal*	*dno*	*Mgro*	*dname*	*dloc*
101	Ram	1000	10		Research	Delhi
102	Shyam	2000	10	101	Research	Delhi
103	Rahul	3000	20	101	Sales	Kolkata

Fig. 4.8.

Select *

From <table> [LEFT/RIGHT/FULL] Join <table 2>; is used for left outer rioght outer, full outer join,

4.10. SUBQUERIES & CORRELATED QUERIES

Till now only simple comparisons conditions is used in where clause. It is possible to have a query result in the condition of a where clause in such case the query in where clause is called **subquery** & complete select statement is called **nested query**.

i.e., a subquery is a query inside a query. The subquery is executed first & its output is used as input for the outer query.

A subquery is on the right side of compression & can return one or multiple values.

Example 4.16. List the employees who are earning more than the average salary of all the employees.

Solution.

```
Select          *
from            employee
Where           sal
(select AVG)    (sal) from employee);
```

First subquery is executed in order to find the average salary of all the employees, then the result of subquery is used in comparison of where clause.

Example 4.17. List all employees who are working in a department located in 'Delhi'.

Solution.

```
Select      *
from        employee
Where       deptno IN (select deptno from department where
            dloc = 'Delhi});
```

4.10.1. Any, All Clause

In matches only those rows that are equal to at least one of the values in the result of subquery. If you have to use inequality comparison. Any & All are used.

Example 4.18. Find employees who are not working in departno = 10 & earn more than all employees working in deptno = 10.

Solution.

```
Select      *
From        employees
Where       Sal > All (select sal from employe where deptno =
            10)
AND deptno ! = 10;
```

Only theose tuples are selected in which sal is greater than all the values return by subquery.

Example 4.19. Find the employees who works in department 10 & earn at least as much any employee working in department 20

Solution.

```
Select    *
from      employee
Where     sal > any (select sal from employee where deptno =
          20)
AND deptno = 10;
```

4.10.2. Correlated SubQuery

Unlike subquery, correlated subquery is a subquery that executes once for each row in the outer query.

i.e., the inner query is related to outer query by having a reference to a column of the outer query.

Example 4.20. Find name of all employees who are working in the same department as their manager.

Solution.

Select Name
from employee E_1
Where deptno IN (select deptno from employee E_2 where E_2. eno = E_1, mgrno);

here subquery is having a reference to a column of other query.

4.11. ORACLE DATA DICTIONARY

Orcale data dictionary is the central part of oracle and it contains the meta data *i.e.*, it contains all the information about the structure and objects of the database such as tables, their attributes, users, constraints etc. all such information is stored in tabllar form. DBA has full access right to data dictionary.

Data dictionary tables :

The installation of oracle creates 3 standard users.

1. SYS : is the owner of all data dictionary and views and has highest previleges to manage objects and structures of oracle database.
2. SYSTEM : is the owner of tables used by different tools such as SQL *FORMS, SQL* REPORTS, has less previleges than sys.
3. PUBLIC : dummy user. All previleges assigned to this user are automatically assigned to all users known in the database.

The command

SELECT * from DICT [IONARY];

List all the tables and views of the data dictionary that are accessible to the user. It includes the name and short description of each Table & view.

SELECT * FROM TAB;

Gives name of all tables owned by the user who has given the command. Data dictionary consists of 2 level.

(A) Internal level contains all base Tables that are used by the various DBMS software components and they are normally not accessible by end users.

(B) External level provides various views on these base tables to access information about objects and structures at different level. This view provided by dictionary are divided into 3 groups USER, ALL, DBA. The group name build prefix for each view name.

1. USER : Tuples in the USER views contains information about objects owned by the account performing the SQL query.

USER_TABLES : all the tables with their name, number of columns, storage inforamtion, slatestical information etc (TABS)

USER_CATALOG : Tables, view, synonyms (CAT)

USER_COL_COMMENTS; comments on columns

USER_CONSTRAINTS : constraint definitions for tables.

USER_INDEXES : all inforamtion about indexes created for tables (IND)

USER_OBJECTS : all database objects owned by user (OBJ)

USER_TAB_COLUMNS : columns of the table & views owned by user (COLS)

USER_TAB_COMMENTS : comments on tables & views

USER_TRIGGERS : Triggers defined by the user

USER_USERS : information about current user

USER_VIEWS L views defined by the user

2. ALL : Rows in the ALL views include rows of the USER view and all inforamtion about objects that are accessible to the current user.

ALL_CATALOG

ALL_TABLES

ALL_OBJECTS

ALL_TRIGGERS

ALL_USERS

ALL_VIEWS

3. DBA : The DBA views consists information about all database objects, regardless of the owner. Only users with DBA privilege can users these views.

DBA_TABLES, DBA_CATALOG, DBA_OBJECTS, DBA_DATA_FILES, DBA_USERS.

Summary

- SQL is used to interact with RDBMS.
- 4 components of SQL are there namely DDL, DML, DQL & DCL.
- You can join tables using inner join and OUTER join.
- Inner join may theta join, natural join or equi join.
- OUTER join may be left, right or full outer join.
- Aggregate function such as AVG, SUM etc can be applied either on entire table or on the group based on some attribute.
- Subqueries & correlated queries are used where it is necessary to process data based on other processed data.
- SQL is used to interact with RDBMS and Oracle in a RDBMS so oracle provides SQL plus which is an extension to the standard SQL.

- It has an on line command interpreter. User can create programs using 4 components of SQL namely DDL, DCL, DML and DQL. User can join more than 2 tables using INNER join and OUTER join.
- Inner join may be theta, natural or equi join where as outer join may be left, right or full outer join.
- Inner join may be theta, natural or equi join whereas outer join may be left, right or full outer join. Aggregate functions such as AVG, MAX COUNT can also be applied either on entire table or on group formed based on some conditions.
- SQL has certain limitations that it doesnot provides programming capabilities likes loops, if constructs.

Multiple Choice Questions

1. SQL stands for
(*a*) Structured query language (*b*) Simple query language
(*c*) Structural questioning language (*d*) None

2. DCL stands for
(*a*) Data calculation language (*b*) Data control language
(*c*) Data communication language (*d*) none

3. INSERT and update are
(*a*) DDL (*b*) DCL
(*c*) DML (*d*) none

4. Grant and Revoke are
(*a*) DDL (*b*) DML
(*c*) DCL (*d*) none

5. Oracle provides a special table that can be used to test any function.
(*a*) DUAL table (*b*) Salary table
(*c*) Employee table (*d*) Sys table

6. DBA stands for
(*a*) data base administrator (*b*) Database accumulator
(*c*) data base animator (*d*) none

7. A query wtih in a query is called as
(*a*) New query (*b*) Child query
(*c*) Sub query (*d*) none

8. ODBC stands for
(*a*) Object data base connectivity (*b*) Online data base connection
(*c*) Open data base connectivity (*d*) none

9. Theta join is a
(*a*) inner join (*b*) outer join
(*c*) full outer join (*d*) none

10. % and – (underscore) are
(*a*) like operators (*b*) arithmetic operators
(*c*) aggregate function (*d*) relational operators

11. ALTER command can be used to

(*a*) change the table structure
(*b*) change the table data
(*c*) add rows to the table
(*d*) delete rows from table

12. The result of a SQL SELECT statement is a

(*a*) report
(*b*) form
(*c*) file
(*d*) relation

13. Keyword BETWEEN is used :

(*a*) for ranges
(*b*) to limit the columns shown
(*c*) as a wildcard
(*d*) None of the above

14. A subquery must be

(*a*) in braces
(*b*) in CAPITAL LETTERS
(*c*) in paranthesis
(*d*) in brackets

15. to remove duplicate rows, the qualifier used :

(*a*) ONLY
(*b*) UNIQUE
(*c*) DISTINCT
(*d*) SINGLE

Selected Exercises

1. What is the difference between CHAR & VARCHAR 2 ?

2. What is view ? How it is different from table ?

3. What is a correlated query how it is different from rested query ?

4. Write a SQL query that gives the second highest salary paid to employees.

5. What is a view? How it is different from a table?

6. What is a self join? Give an example to explain. Self joins.

7. Explain various data types, available in oracle.

8. The oracle saver performs certain steps when having clause in used. What are those steps?

9. What s the difference between truncate table and delete table in SQL.

10. What are the various characteristics of SQL? Discuss five aggregate functions with suitable examples?

11. Consider the following 2 tables.

A	B	C
10	*a*	5
15	*b*	8
25	*a*	6

A	B	C
10	*b*	6
25	*c*	3
10	*b*	5
	a	7

(*a*) Give the result of join of these 2 Tables

(*b*) Is there any way to find the primary key of the above tables.

12. Let *R*(*A*, *B*, *C*) be a relation schema and let *r*, and r_2 both be the relations on *R*. Give SQL equivalent of the following:

(*a*) $r_1 \cup r_2$
(*b*) $r_1 \cap r_2$
(*c*) $r_1 - r_2$
(*d*) $\pi_{AB}(r_1) \bowtie \pi_{BC}(r_2)$

13. List the following relation scheme be given

Employee (SSN, name, age d no)

Salary (SSN, Salary)

Work on (project#, SSN)

Project (Project#, Project name, Loc)

Write queries in SQL

(*i*) display the names of projects at 'Delhi'

(*ii*) find the project name of employee whose salary is greater than 2000.

(*iii*) Retrieve the name and SSN of employees working on project 'Reseach'

14. Consider the following relations

Employee (emp name, street, city)

Works (emp name, company name, salary)

Company (company name, city)

Manages (emp name, manager name)

Write down the following quicks in SQL

(*a*) find the name of all employee who live in the same city on do their managers.

(*b*) find the name of city where employee '*X*' works

(*c*) find the name and cities of residnece for all employee who work for company "*XYZ*".

❑❑❑

Chapter

5 Introduction to PL/SQL

Inside this Chapter

5.1. PL/SQL

5.1.1. Introduction

The development of database applications typically requires language constructs similar to those that can be found in programming languages such as C, C++, or Pascal. These constructs are necessary in order to implement complex data structures and algorithms. A major restriction of the database language SQL, however, is that many tasks cannot be accomplished by using only the provided language elements.

PL/SQL (Procedural Language/SQL) is a procedural extension of Oracle-SQL that offers language constructs similar to those in imperative programming languages. PL/SQL allows users and designers to develop complex database applications that require the usage of control structures and procedural elements such as procedures, functions, and modules.

The basic construct in PL/SQL is a block. Blocks allow designers to combine logically related (SQL-) statements into units. In a block, constants and variables can be declared, and variables can be used to store query results. Statements in a PL/SQL block include SQL statements, control structures (loops), condition statements (if-then-else), exception handling, and calls of other PL/SQL blocks.

PL/SQL blocks that specify procedures and functions can be grouped into **packages**. A package is similar to a module and has an interface and an implementation part. Oracle offers several predefined packages, for example, input/output routines, file handling, job scheduling etc. (see directory $ORACLE_HOME/rdbms/admin).

Another important feature of PL/SQL is that it offers a mechanism to process query results in a tuple-oriented way, that is, one tuple at a time. For this, **cursors** are used. A cursor basically is a pointer to a query result and is used to read attribute values of selected tuples into variables. A cursor typically is used in combination with a loop construct such that each tuple read by the cursor can be processed individually.

In summary, the major goals of PL/SQL are to

- increase the expressiveness of SQL,
- process query results in a tuple-oriented way,
- optimize combined SQL statements,
- develop modular database application programs,
- reuse program code, and
- reduce the cost for maintaining and changing applications.

5.2. INTRODUCTION TO PL/SQL

Procedural language/structured query language is a procedural extension of oracle-*SQL*. *SQL* has limitations such that it does not offer programming language constructs such as loops, control statements, variables, error management etc. *PL/SQL* offers all such constructs. It is block structured language. Each block builds a program unit and block can be nested_user can use *PL/SQL* to create the following :

- anonymous *PL/SQL* blocks
- triggers
- cursors
- stored produces
- *PL/SQL* functions

Blocks that specify procedures and functions can be grouped into packages.

General structures of a *PL/SQL* block

```
[<Block header>]
[DECLARE
        <constants>
        <variables>
        <cursors>
        <user defined exceptions>]
BEGIN
        <PL/SQL statements>
        [EXCEPTION
                <Exception bandling>]
END;
```

Key Points

1. Block header specifies whether the block is a procedure, a function, or a package, if no header is specified than block is said to be an **anonymous** *PL/SQL* block.
2. Consants, variables, cursors, exceptions used in a block must be declared in the DECLARE section of that block.

   ```
   <variable name> [CONSTANT] <date type> [NOT NULL] [:= <Expression>];
   ```

Sample DECLARE

dt	DATE;	there 3 variables
st	VARCHAR (8) := 'Salesman';	namely dt, st, sn
sm	CONSTANT number (3, 2) : = 1.5;	are declared.

3. Instead of specifying a data type, we can also refer to the data type of a table's column.

example Employee . eno % TYPE;

Table name Column name

4. We can also declare a variable that can store a complete tuple from a given table.

   example DEPT % ROW TYPE;
5. Exceptions are used to process errors and warnings that occur during execution.
6. Data type can be CHAR, VARCHAR 2, NUMBER, DATE and BOOLEAN.

5.3. LANGUAGE ELEMENTS

5.3.1. While Loop

```
While <condition> LOOP
        PL/SQL statements
END LOOP;
```

5.3.2. For Loop

```
For <index> IN [REVERSE] <lower bound> .. <upper bound>
Loop
            <PL/SQL>
END LOOP;
```

5.3.3. If Statement

```
If <condition> THEN <sequence of statements>
[ELSIF] <condition> THEN <sequence of statements>
..........
..........
[ELSE] <sequence of statements>
ENDIF;
```

NOTE :
1. Type SET SERVER OUTPUT ON to enable the client console to receive messages from the server side.
2. To send messages from the *PL/SQL* block use DBMS_OUTPUT . PUT-LINE ('Hi');

Example 5.1. Write a *PL/SQL* code that prompts user to enter the account number, check if the user account balance is less than the minimum balance, deduct Rs. 100 from the balance.

Solution.

```
DECLARE
        Curbal      number (10, 2);
        Curacno     varchar (7);
        fine        number (4) : = 100;
        minbal      CONSTANT NUMBER (7, 2) : = 1000.00;
```

```
BEGIN     Curacno : = & curacno ;
          Select     bal INTO curbal
          from       account
          where      acno = curacno;
          if         curbal < minbal
THEM
          UPDATE     account
          SET        bal = bal-fine
          WHERE      acno = curacno;
ENDIF;
END;
```

5.4. CURSORS

A cursor is a special construct to hold the data rows returned by *SQL* query. There are 2 types of cursors implicit cursor and explicit cursor.

Implicit cursor is automatically created when the *SQL* statement returns only one value. Whereas the explicit cursor is created to hold the output of *SQL* statement that may return 2 or more rows.

Explicit cursor has following steps :

1. first cursor is declared
2. open the cursor
3. fetch tuples
4. close the cursor

1. CURSOR <cursor name> [<list of parameter>] is <select statement>;
2. OPEN <cursor name> [<list of parameter>];

   Opening the cursor executes the *SQL* command and populates the cursor with data and open the cursor for processing.
3. FETCH <cursor name> INTO <list of variables>;

   Once the cursor is opened, you can fetch the data from the cursor into *PL/SQL* variables.
4. CLOSE <cursor name>;

   After all processing is done finally close the cursor.

e.g.

```
DECLARE
CURSOR      Emp_cur IS      SELECT *from employee;
            Emp_record      Employee % ROW TYPE;
            Emp_sal         Employee sal % TYPE
BEGIN
            OPEN  Emp-cur;
            LOOP
                  FETCH Emp_cur INTO emp_record;
                  EXIT WHEN
                            Emp_cur% NOT FOUND;
            Emp_sal : = Emp_record. sal;
```

```
        <sequence of statements>
        END LOOP;
        CLOSE           Emp_cur;
END;
```

The above cursor is used find the salaries of all employees.

Example 5.2. Find job title, name of employees who have been hired after a given date and who have a manager working in given department from employee.

Solution. DECLARE

```
        CURSOR Emp_sec (startdate DATE, dno NUMBER) is
                SELECT  job, name
                from    Employee E
                WHERE   hiredate > startdate
                AND     EXISTS (SELECT * from Employee E1
                        WHERE E. mgrno = E1. Eno AND E1. dno = dno);
        BEGIN
        .........
        .........
        END;
```

5.5. TRIGGERS

A trigger is a *PL/SQL* code that is executed automatically by *RDBMS* as a side effect of database modifications. Triggers are used for maintaining the data integrity and consistency as well as to enforce business logics.

A trigger is referred to as the event-condition-action model because trigger has 2 requirements

(*a*) specifies when a trigger is to be executed.

(*b*) specifies the actions to be taken When the trigger executes.

A trigger is invoked before or after a data row (tuple) is inserted, updated, or deleted. Basically a trigger is associated with a table and table may have one or more triggers.

General Syntax

```
CREATE TRIGGER <trigger name>
[BEFORE/AFTER] [DELETE|INSERT|UPDATE OF <Column>] ON <table name>
[FOR EACH ROW]
[DECLARE
        [<variable name> <data type> : = <value>]
BEGIN
        <PL/SQL statements>
END;
```

Trigger events indicates when the trigger is to be executed *i.e.* before or after of insertion, deletion or updation of a tuple.

2 types of triggers are there.

If FOR EACH ROW keyword is not there then it is statement level trigger and is executed only once before ior after the triggering statement is completed (default one) whereas if FOR EACH ROW is there then it is row level trigger and is executed once for each ROW affected by the triggering statement. *i.e.* if you are deleting 10 rows, trigger executed 10 times.

Trigger action s enclosed between BEGIN and END.

Example 5.3. Create a trigger overdraft after the update of an account.

Solution.

```
CREATE TRIGGER overdf
    After UPDATE ON account
    REFERENING NEW ROW as nrow
    FOR EACH ROW
    WHEN        nrow.bal < 0
    BEGIN
        INSERT INTO borrower  (SELECT (name, acno FROM depositor
                                WHERE nrow.acno = depositor. acno);
        INSERT INTO LOAN VALUES (nrow.acno, nrow branch, –nrow. bal);
        UPDATE    account
        SET       bal = 0
        WHERE     account acno = nrow. acno;
    END;
```

Referencing new row as clause creates a variable nrow, which stores the value of an updated row after the update.

The previous trigger checks the account balance after update and if the account balance is negative, is loan is created.

You can delete a trigger by typing

DROP TRIGGER <trigger name>;

5.6. STORED PROCEDURES

Stored procedure is a named collection of procedural and sq. statements. They are stored in a database. They are used to encapsulate and represent business transactions.

```
CREATE [REPLACE] PROCEDURE
        <procedure name> [<list of parameter>] is
                <Declarations>
BEGIN
        <PL/SQL statements>
END;
```

Example 5.4. Write a procedure to increase the salary of all employes who work in the department given by the procedure parameter.

Solution : CREATE PROCEDURE incsal (dn NUMBER, pt NUMBER DEFAULT 0.5) is

```
CURSOR  EMPER (deptno NUMBER) is
        SELECT    sal
        FROM      Employee
        WHERE     dno = deptno
```

```
Empsal    NUMBER (8);
BEGIN
          OPEN          EMPER (dn);
          LOOP
               FETCH    EMPER INTO Empsal;
               EXIT     WHEN EMPER % NOT FOUND;
               UPDATE   Employee
               SET      sal = empsal * [(100 + Pt)/100]
               WHRE     CURRENT OF EMPER;
END LOOP;
CLOSE EMPER;
Commit;
END incsal;
```

To execute the above procedure from *SQL** PLUS EXECUTE incsal (10, 3).

The above stored procedure, we are declaring 2 variables one is the cursor and other is normal variable.

5.7. FUNCTIONS

Stored procedure and functions are similar except function returns a value whose data type must be specified.

```
CREATE FUNCTION <function name> [<List of parameters>]
RETURN      <data type>  [is]
BEGIN
            <sequence of statements>
            RETURN <value/expression>;
END;
```

Example 5.5. Create a function to find out the total salary paid by a department.

Solution.

```
CREATE FUNCTION getsal (dn NUMBER) RETURN NUMBER is total sal NUMBER;
BEGIN
          total sal : = 0
          FOR empsal IN  (SELECT    sal
                          FROM      Employee
                          WHERE     dno = dn.
                          AND       sal is NOT NULL)
          LOOP
                     total_sal : = total_sal + emp_sal. sal;
          END        LOOP;
          RETURN     total_sal;
END getsal;
to execute,          EXECUTE : sl : = getsal (20);
```

Summary

- PL/SQL is a procedural extension of SQL.
- It offers language constructs such as loops, control/statement and exception handling. We can write triggers, stored procedures, functions with PL/SQL.
- Exception handling can be done.
- Business logics are implemented using PL/SQL constructs.
- PL/SQL (procedural language) structured query language is an extension of SQL. It is the supperset of of SQL. PL/SQL offers language constructs such as loops, control statements and exception handling mechanisms. We can say business logics are implemented by PL/SQL code via writing the trigger, stored procedures, functions etc.
- The oracle engine uses a work area to execute SQL statements and store information. We can name this area. PL/SQL provides cursor to name this area. PL/SQL provides cursor to name there work areas and access the data stored into it.

Multiple Choice Questions

1. A buffer use to store results of the recent query.
(*a*) Package (*b*) Trigger
(*c*) Exception (*d*) Cursor

2. The OUT parameter behaves use a
(*a*) Constant (*b*) Uninitialized variable
(*c*) Regular variable (*d*) None

3. Functions returns using the statement
(*a*) Call (*b*) Return
(*c*) Return <datatype> <function name> (*d*) None

4. Gives number of rows retrieved by cursor
(*a*) % FOUND (*b*) % RETURN
(*c*) % FOUND % (*d*) % ROW COUNT

5. Gives whether the rows was retrieved by cursor or not.
(*a*) % ROW COUNT (*b*) % FOUND
(*c*) % NOT FOUND (*d*) % IS OPEN

6. For what purpose are views used ?
(*a*) Hide rows (*b*) Hide SQL statements
(*c*) Hide columns (*d*) all of the above

7. ORACLE supports which kinds of triggers ?
(*a*) INSTEAD of (*b*) BEFORE
(*c*) AFTER (*d*) ALL

8. What is NOT an advantage of stored procedures ?
(*a*) Code sharing (*b*) Security
(*c*) Optimization (*d*) Measured netowrk traffic

9. .... is a PL/SQL block that can defined to automatically execute for insert, update and delete statements against a table
(*a*) Database (*b*) Database trigger
(*c*) Assertion (*d*) None

10. If the entire table is present, and there is no computed values, the view is
(*a*) Updatable (*b*) Not needed
(*c*) Non-updatable (*d*) None

Selected Exercises

1. What is difference between SQL and PL/SQL ?

2. What is a stored procedure ? Why it is particularly useful ? Give an example.

3. What do you mean by % ROW COUNT, % FOUND, % NOT FOUND and % IS OPEN ?

4. How triggers are executed ?

5. Write a procedure to add a new employee to EMPLOYEE table.

6. Write a PL/SQL code to retrive the employer name, age and designation of an employee whose employee is accepted by system.

7. Write a PL/SQL code to insert rows into a new table

8. Write a PL/SQL code that gives second highst salary.

9. Write a PL/SQL code to print the employee number, name and salary of 10 highest paid employees.

10. Write a PL/SQL code to update the salary of employee who earn less than average salary by 10%.

11. What do you mean by trigger? What are its advantages? How triggers can be modified?

12. Write a procedure to increase the salary of all employees who works in the departemnt gives by procedure as parameters and also the increment percentage is given as second parameter to procedure.

13. Create a stored function that accepts two numbers and returns the multiplication of these two numbers.

14. Write a PL/SQL code that accept the department number and calculate the total number of employees in that department.

15. Explain the variable attributes:

(*i*) % TYPE

(*ii*) % ROW TYPE

16. Write a trigger for "every loan has at least one customer who maintains an account with minimum balance of Rs. 1000 in banking system".

17. Create a table to represent saving bank account of a bank consists of accno, custname, balance and write a PL/SQL block to implement deporite and withdraw. Withdrawl should not be allowed if the balance goes below Rs. 500.

❑❑❑

Chapter

6 Normalization

Inside this Chapter

6.0. INTRODUCTION

ER model gives table structure of the database. The table structure given by *ER* model may be or may not be good. *i.e.* the tables created by *ER* model may have redundancies and various anomalies. So how do we recognize a bad database design or good database design ? It is by normalization. Thus Normalization is the process of refining the data structure built by *ER* model. It is first proposed by *CODD* (1972).

Normalization takes a relation scheme through a series of tests to certify whether it satisfy a certain normal form. It works through a series of stages called NORMAL FORMS. *e.g.* FIRST NORMAL FORM (1 *NF*), SECOND NORMAL FORM (2 *NF*), THIRD NORMAL FORM (3 *NF*), CODD proposed all 1 *NF*, 2 *NF*, 3 *NF*, latter on Boyee and CODD proposed BCNF and FOURTH NORMAL FORM (4 *NF*) and FIFTH NORMAL FORM (5 *NF*).

2 *NF* is better than 1 *NF* and 3 *NF* is better than 2 *NF* but you should not assume that the highest level normal form is always the most desirable. Generally the higher the normal form, the more relational JOIN operations are required to produce a specific output *i.e.* time consuming approach.

Thus the normalization is a process of analyzing the given relation schema based on their functional dependencies and primary key to achieve the following

1. Minimizing the redundancy.
2. Removing the insertion, deletion and updation anomalies.

Bad design ⟶ | Normalization | ⟶ Good design

Fig. 6.1.

What you do in Normalization ?

Unsatisfactory relation scheme that do not meet certain conditions—the normal form test – are decomposed into smaller relation schemes that satisfy these test.

2 properties must be satisfied when decomposition take place

1. the decomposition must be lossless join decomposition or non additive join.
2. the dependency preservation property, which ensure that each functional dependency is there in some individual relation scheme after decomposition.

6.1. FUNCTIONAL DEPENDENCY

The concept of functional dependency is very important. Let X, Y be the two subsets of reltion schema R then if the given value of X, there is only one value of Y, then Y is said to functionally dependent on X or X functionally determines Y. We write

$$X \longrightarrow Y$$

i.e. functional dependency is a constraint between two subsets of attributes of relation schema R.

The constrain is that, for any two tuples t_1, t_2 in $r(R)$ that have

$$t_1[X] = t_2[X] \text{ we must also have}$$
$$t_1[Y] = t_2[Y]$$

- If $X \rightarrow Y$ hold in R, we can not say that whether $Y \rightarrow X$ holds in R on not.
- If there can not be more than one tuple with a given X value in r then X is a candidate key of R.
- A functional dependency $X \rightarrow Y$ is trivial if $X \supseteq Y$ and all other functional dependencies are not trivial.

6.1.1. Inference Rule for Functional Dependency

Let F be the given set of functional dependencies specified on schema R. We can infer or deduce other functional dependencies from F by applying some rules. The set of all such functional dependencies, inferred by F, is called as closure of F and is denoted by F^+.

Rules are called as Armstrong Rules.

R_1 : Reflexive Rule :

If α is a set of attribute such that $\beta \subseteq \alpha$ then $\alpha \rightarrow \beta$.

R_2 : Augumentation Rule :

If $\alpha \rightarrow \beta$ holds in R and Y is a set of attribute from R then $\gamma\alpha \rightarrow \beta\gamma$ holds.

R_3 : Transitive Rule :

If $\alpha \rightarrow \beta$ and $\beta \rightarrow \gamma$ holds in R then $\alpha \rightarrow \gamma$ also holds.

R_4 : Union or Additive Rule :

If $\alpha \rightarrow \beta$ and $\alpha \rightarrow \gamma$ hold in R then $\alpha \rightarrow \beta\gamma$ holds in R.

R_5 : Decomposition or Projective Rule :

If $\alpha \rightarrow \beta\gamma$ hold in R then $\alpha \rightarrow \beta$ or $\alpha \rightarrow \gamma$ holds in R.

R_6 : Pseudotransitive Rule :

If $\alpha \rightarrow \beta$ and $\gamma\beta \rightarrow \delta$ holds in R then $\gamma\alpha \rightarrow \delta$ holds in R.

Example 6.1. Given a set of *F.D.F* as $\{A \rightarrow B, ABCD \rightarrow E, EF \rightarrow G\}$ then find whether $ACDF \rightarrow G$ is implied by the set *F*.

Solution : Given $A \rightarrow B$, $ABCD \rightarrow E$

Then, $ACD \rightarrow E$, by pseudotransitivity

$ACDF \rightarrow EF$, by Augumentation rule

$ACDF \rightarrow G$, by transitivity

Hence proved.

Example 6.2. Given R (A, B, C, G, H, I) and $F = \{A \rightarrow B, A \rightarrow C, CG \rightarrow H, CG \rightarrow I, B \rightarrow H\}$ is $AG \rightarrow I$ holds in R ?

Solution. Given $A \rightarrow C$, $CG \rightarrow I$

Then $AG \rightarrow I$, by pseudotransitivity

Hence proved.

6.1.2. Procedure to Compute Closure of F using Armstrong Rules

```
F+ = F
Repeat
        for each functional dependency in F+
            use reflexive rule and augumentation rule on F
            add the resulting functional dependency to F+
        for each pair of f1 and f2 in F+
        find whether transitivity is there or not
        if yes then add the resulting
        functional depending to F+
Until F+ does not change.
```

6.1.3. Closure of an Attribute Set

Let α be the set of attributes of Relation schema R. Then set of all attributes functionally determined by α, denoted by α^+, under the given set of functional dependencies is called the closure of attribute set α.

Algorithm to find closure of attribute set

$\alpha^+ = \alpha$

While change to α^+ do

for each functional dependency $X \rightarrow Y$ in F

do

Begin if $X \subseteq \alpha^+$ then

$$\alpha^+ = \alpha^+ \cup Y$$

END

Example 6.3. Given R = ($ABCGHI$) and set of functional dependency $F = \{A \rightarrow B, A \rightarrow C, CG \rightarrow H, CG \rightarrow I, B \rightarrow H\}$ compute $(AG)^+$?

Solution. According to algorithm $\alpha = AG$

Then $\alpha^+ = \{AG\}$

for $A \rightarrow B$, $\alpha^+ = \{A, B, G\}$ as A is a subset of $\{AG\}$

for $A \rightarrow C$, $\alpha^+ = \{A, B, C, G\}$ as $A \subseteq \{A, B, G\}$

for $CG \rightarrow H$, $\alpha^+ = \{A, B, C, G, H\}$ as $\{C, G\} \subseteq \{A, B, C, G\}$

for $CG \rightarrow I$, $\alpha^+ = \{A, B, C, G, H, I\}$ as $\{C, G\} \subseteq \{A, B, C, G, H\}$

for $B \rightarrow H$, $\alpha^+ = \{A, B, C, G, H, I\}$ so $\{A, G\}^+ = \{A, B, C, G, H, I\}$

Example 6.4. Given R ($ABCDEFG$) and set of functional dependencies ? $F = \{A \rightarrow B, BC \rightarrow DE, AEF \rightarrow G\}$. Then find $\{A, C\}^+$

Solution. $\alpha^+ = \{A, C\}$

for $A \rightarrow B$, $\alpha^+ = \{A, B, C\}$ as $A \subseteq \{A, C\}$

for $BC \rightarrow DE$, $\alpha^+ = \{A, B, C, D, E\}$ as $BC \subseteq \{A, B, C\}$

for $AEF \rightarrow G$, $\alpha^+ = \{A, B, C, D, E\}$ as $AEF \not\subset \{A, B, C, D, E\}$

Hence $\{A, C\}^+ = \{A, B, C, D, E\}$.

Example 6.5. Given $A \rightarrow BC$, $E \rightarrow CF$, $B \rightarrow E$ and $C \rightarrow EF$ compute $\{A, B\}^+$?

Solution. $\alpha^+ = \{A, B\}$

for $A \rightarrow BC$, $\alpha^+ = \{A, B, C\}$ as $A \subseteq \{A, B\}$

for $E \rightarrow CF$, $\alpha^+ = \{A, B, C\}$ as $E \not\subseteq \{A, B, C\}$

for $B \rightarrow E$, $\alpha^+ = \{A, B, C, E\}$ as $B \subseteq \{A, B, C\}$

for $C \rightarrow EF$, $\alpha^+ = \{A, B, C, E, F\}$ as $C \subseteq \{A, B, C, E\}$

Hence, $\{A, B\}^+ = \{A, B, C, E, F\}$.

There are several use of attribute closure.

1. to test whether a is a super key of R, compute α^+ if α^+ contains all attribute of R then α is a super key of R.
2. to check a *fd* $\alpha \rightarrow \beta$ holds or not. If $\beta \subseteq \alpha^+$ then by refexivity $\alpha \rightarrow \beta$ holds in R.

6.1.4. To find Super Keys, Candidate Keys and Irreducible

Example 6.6. Given $F = \{ABCD \rightarrow E, AB \rightarrow G, B \rightarrow F, C \rightarrow J, CJ \rightarrow I, G \rightarrow H\}$ find super key, candidate key and irreducible.

Solution. By the left hand side of all functional dependencies given in the set F,

the super key $= \{A, B, C, D, G, J\}$

Now, to find candidate key from the above super key we will eliminate those attributes form super key which can be functionally determined by any attribute(s) from super key.

So, $\{A, B, C, D, G, J\}$

$= \{A, B, C, D, G\}$ we eliminate J because given $C \rightarrow J$

$= \{A, B, C, D\}$ because $AB \rightarrow G$ so eliminate G

Now $\{A, B, C, D\}$ is not functionally determined by any functional dependency of F. So irreducible set of F is $\{A, B, C, D\}$ and hence the candidate key.

Example 6.7. Given $F = \{A \rightarrow B, BC \rightarrow D, BC \rightarrow E, AEF \rightarrow G, B \rightarrow G\}$ find candidate key.

Solution. Super key $= \{A, B, C, , E, F\}$, by the left hand side of all f.d.s

$= \{A, C, E, F\}$ as $A \rightarrow B$ so eliminate **B**

$= \{A, C, F\}$ as $A \rightarrow B$, $BC \rightarrow E$ so by pseudo $AC \rightarrow E$

So candidate key $= \{A, C, F\}$.

Example 6.8. Given $F = \{A \rightarrow B, B \rightarrow CD, C \rightarrow E, D \rightarrow C, E \rightarrow F\}$ find the candidate key.

Solution. Super key $= \{A, B, C, D, E\}$, by the left hand side of all f.d in F

$= \{A, C, D, E\}$ as $A \rightarrow B$, eliminate B

$= \{A, C, D\}$ as $C \to E$ eliminate E
$= \{A, D\}$ as $D \to C$ eliminate C

The candidate key is $\{A, D\}$

Now, again super key $= \{A, B, C, D, E\}$
$= \{A, B, E\}$ as $B \to CD$ eliminate CD
$= \{A, E\}$ as $A \to B$ eliminate B

candidate key $= \{A, E\}$

Therefore 2 candidate keys $\{A, D\}$ and $\{A, E\}$.

Example 6.9. Given $F = \{AB \to C, A \to DE, B \to F, F \to GH, D \to IJ\}$ find candidate key.

Solution. Super key $= \{A, B, D, F\}$
$= \{A, B, D\}$ as $B \to F$

Candidate key $= \{A, B, D\}$.

Example 6.10. Given $\{A \to BC, B \to E, CD \to EF\}$ find the candidate key.

Solution. Super key $= \{A, B, C, D\}$
$= \{A, D\}$ as $A \to BC$

Candidate key is $\{A, D\}$.

6.2. PROBLEMS WITH UNNORMALIZED DATABASE

The database that is not normalized may suffers from the following problems.

1. Data redundancy *i.e.* same information is stored at various places. We know that data redundance leads to data inconsistency, more storage and extra cost.
2. Various anomalies like insertion, deletion and updation anomalies are there.

Let us consider an example database to understand the above problems. A company is managing several projects it has to maintain information regarding the each project such as its name, project number, employees working in the project along with the personal information of all employees such as their name, job.

The company charge its client by billing/day spent on each contract. The daily billing is different for different employee.

One can design the database for the company by having a single table company having all the above information.

COMPANY (p#, pname, empno, ename, job, chg/day, nday). Any moment, the database state may be ;

p#	*pname*	*empno*	*ename*	*job*	*ch/day*	*nday*
1.	Insurance	101	Rahul	Accountant	1000	30
		103	Amit	Manager	4000	15
		105	Rakesh	Clerk	500	30
		106	Rohit	Clerk	500	25
2.	Health	101	Rahul	Accountant	1000	30
		103	Amit	Manager	4000	10
		107	Ruch	HR	500	30
		109	Suruchi	Clerk	500	30

3.	Environment	105 103 108	Asim Amit Ravi	Accountant Manager Clerk	1000 4000 500	30 15 30
4.	Public relation	104	Ruchit	Client support	1000	30

Fig. 6.2. The database state of COMPANY databse **COMPANY_INF**

The following problems are there with above data base

1. Data is redundant in the above table because 'Amit' is manager charging Rs. 4000/day and working in all the 3 projects. By having the above design the information regarding 'Amit' is stored three times.

2. **Anomalies :**

***(i)* Insertion anomaly :** Clearly {*p*#, empno} is the primary key of the above relation. Now, if we have to start a new project say 'water sanitation' having no employee assigned to it the tuple should be

<4, 'water sanitation', null, null, null, null>

But the definition of primary key says it can not be null. Thus we will not be able to insert this new tuple.

***(ii)* Deletion anomaly :** Suppose a employees 'Ruchit' leaves the company. Then we have to delete the tuple corresponding to this employee. When we do this, the information regarding the project **'public relation'** get deleted because 'Rachit' was the only employee working in that project.

***(iii)* Updation anomaly :** Suppose we change the job of 'Amit' from manager to General Manager for the project insurance, then we have to change all the occuurence of 'Amit' otherwise 'Amit' will have 2 job descriptions.

Because of all the above anomalies and data redundancy, normalization process is carried out.

6.3. NORMAL FORMS

6.3.1. First Normal Form

A relation is in 1 *NF* if the domain of each attribute of the given relation is atomic and each attribute takes only a single value from that domain *i.e.* it is not a set of values, list of vlues nor a tuple of values.

In other words, *A* relation *R* is in 1 *NF* if

1. There are no repeating groups in the table.
2. Key must be identified.
3. All attributes are dependent on the primary key.

Company table is not in 1 *NF*, as *p*#, pname repeats for several tuples.

p#	*pname*	*empno*	*ename*	*job*	*ch/day*	*nday*
1	Insurance	101	Rahul	Accountant	1000	30
1	Insurance	103	Amit	Manager	4000	15
1	Insurance	105	Rakesh	Clerk	500	30
1	Insurance	106	Rohit	Clerk	500	25

2	Health	101	Rahul	Accountant	1000	30
2	Health	103	Amit	Manager	4000	10
2	Health	107	Ruchi	HR	500	30
2	Health	109	Suruchi	Clerk	500	30
3	Environment	105	Asim	Accountant	1000	30
3	Environment	103	Amit	Manager	4000	15
3	Environment	108	Ravi	Clerk	500	30
4	Public relation	104	Suchit	Client/support	1000	30

Fig. 6.3. Relations is in 1 NF

Dependency diagram of the above relation is

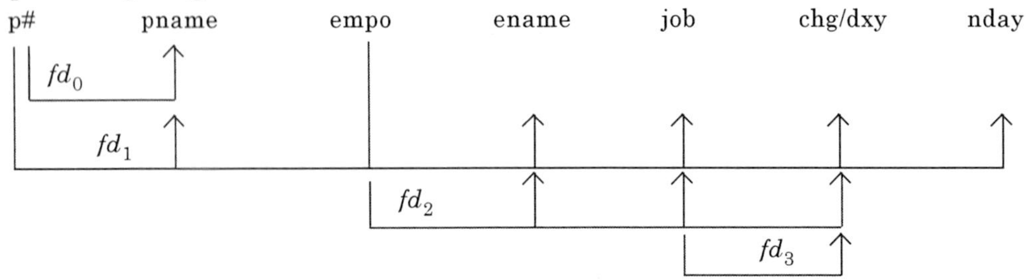

The primary key is {*p*#, Empno}

6.3.2. Second Normal Form

2 *NF* is based on the concept of full functional dependency. A functional dependency may be full or partial. A functional dependency $X \rightarrow Y$, of a relation schema *R*, is said to be a full functional dependency if removal of any attribute *A* from *X* means that dependency does not hold any more.

i.e. $\{X - \{A\}\} \not\rightarrow Y$, where $A \in X$

whereas functional dependency $X \rightarrow Y$, of a relation schema *R*, is said to a partial dependency if some attribute *A* can be removed from *X* and dependency still holds in *R*.

i.e. $\{X - \{A\}\} \rightarrow Y$.

A relation is in 2 *NF* if it is in 1 *NF* and it has no partial dependency *i.e.* no attribute is dependent on only a portion of primary key.

2 *NF* tests only for these functional dependencies whose L.H.S. are part of primary key. There are 2 situations

1. If primary key contains only a single attribute then no need to test for 2 *NF*, relation is in 2 *NF*.
2. If primary key consists of more than one attributes then relation is in 2 NF if every non prime attribute *A* of *R* is fully functionally dependent to the primary key of *R*.

[**Note :** A attribute *A* of *R* is said to be prime attribute if it part of some candidate key of *R* otherwise it is said to be non prime attribute.]

The relation company_1 *NF* is not in 2 *NF* because 2 partial dependencies exists in it

$$p\# \rightarrow pname$$

$$\text{and } empno \rightarrow \{ename, job, chg/day\}$$

i.e. non-prime attributes are partially dependent on the primary key. So we have to decompose the original relation.

Steps to decompose company_1 *NF* are :

1. Write L.H.S. of each partial dependency on a separate line and then write the original primary key of company_1 *NF*.

   i.e. *p#*
   empno
   p#, empno

   it indicates the original relation company_1 *NF* is divided into 3 relations and each relation will have the above attributes as primary key respectively, we can name these relations.

   e.g. PROJECT, EMPLOYEES, WORKS.

2. Identifies all the dependent attributes of 3 keys. From the dependency diagrams and add to them to 3 relations.

PROJECT (**p#,** pname)

Employee (**empno,** ename, job, chg/day)

Works (**p#, empno,** nday) the above decomposed relations are in 2 *NF* the dependency. Diagram is as follows :

PROJECT (p#, pname)

EMPLOYEE (empno, ename, job, chg/day)

WORKS (p#, empno, nday)

PROJECT

p#	*pname*
1	Insurance
2	Health
3	Environment
4	Public relation

EMPLOYEE

Empno	*ename*	*job*	*chg/day*
101	Rahul	Accountant	1000
103	Amit	Manager	4000
104	Ruchit	Client support	1000
105	Rakesh	Clerk	500
106	Rohit	Clerk	500
107	Ruchi	HR	500
108	Ravi	Clerk	500
109	Suruchi	Clerk	500

WORKS

p#	*empno*	*nday*
1	101	30
1	103	15
1	105	30
1	106	25
2	101	30
2	103	10
2	107	30
2	109	30
3	105	30
3	103	15
3	108	30
4	104	30

Fig. 6.4. Relations are in 2 NF

6.3.3. Third Normal Form

3 *NF* is based on the concept of transitive dependency. According to CODD's definition a relation *R* is in 3 *NF* if it is in 2 *NF* and no non prime attribute of *R* is transitively dependent on the primary key of *R*.

The relation scheme Employee is in 2 *NF* but not in 3 *NF* because

EMPLOYEE (empno, ename, job, chg/day)

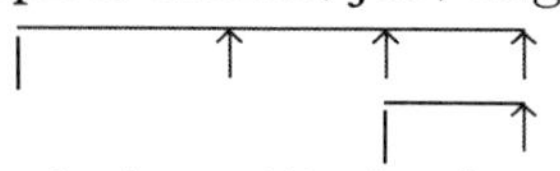

The non prime attribute chg/day is transitively dependent on the primary key (empno) as

empno → job
job → chg/day
and also empno → chg/day.

The employee scheme is to be decomposed, if the relation scheme *R* has 2 transitive dependencies then it is to be decomposed into 3 relations.

Step 1 : For every transitive dependency, write the attribute whose value determines other attribute in a row at primary key of new relation, in the above case only one transitive dependency is there os JOB.

Step 2 : Identifies all the dependent attributes of this new primary key. Put all of them along with new primary key in a separate relation.

JOB-CLASS (**job,** chg/day)

Step 3 : Eliminate all the dependent attributes in the transitive dependencies.

So EMPLOYEE 1 (empno, ename, job)

Thus the employee is broken into 2 relations

EMPLOYEE 1 (**Empno,** ename, job)

JOB-CLASS (**job,** chg/day)

are in 3 *NF*.

Thus,

PROJECT

p#	*pname*
1	Insurance
2	Health
3	Environment
4	Public relation

EMPLOYEE

Empno	*ename*	*job*
101	Rahul	Accountant
103	Amit	Manager
104	Ruchit	Client support
105	Rakesh	Clerk
106	Rohit	Clerk
107	Ruchi	HR
108	Ravi	Clerk
109	Suruchi	Clerk

WORKS

p#	**Empno**	*nday*
1	101	30
1	103	15
1	105	30
1	106	25
2	101	30
2	103	10
2	107	30
2	109	30
3	105	30
3	103	15
3	108	30
4	104	30

JOB-CLASS

job	*chg/day*
Accountant	1000
Manager	4000
Client support	1000
Clerk	500
HR	500

Fig. 6.5. 3 NF

The general definition of 3 *NF* is

A relation scheme *R* is in 3 *NF* if, however a non-trivial functional dependency $X \rightarrow Y$ holds in *R* then either.

(*a*) *X* is a super key of *R*

or

(*b*) *Y* is a prime attribute.

6.3.4. BCNF Boyce-Codd Normal Form

BCNF is much stricter than 3 NF because every relation that is in BCNF is also in 3 NF however, a 3 NF relation not necessarily in BCNF.

When a relation has only one candidate key then 3 NF and BCNF are equal. where as when relation has more than one candidate BCNF can be violated. A relation is in BCNF when every determinant is a candidate key.

Example :

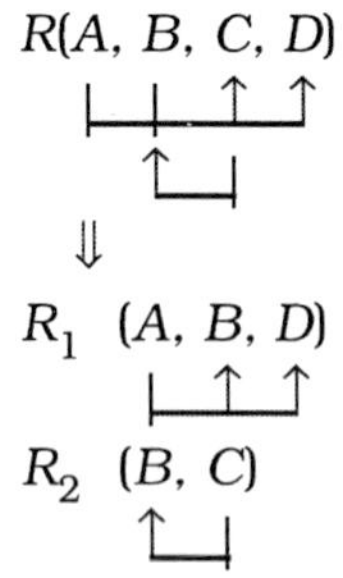

R is in 3 *NF* but not in *BCNF*

General definition : A relation scheme *R* is in *BCNF* if, whenever a non-trivial functional dependency $X \rightarrow Y$ holds in *R*, then *X* is a super key of *R*.

6.3.5. 4 NF (Fourth Normal Form)

All of previous normal forms, (1 *NF*, 2 *NF*, 3 *NF*, *BCNF*) were based on the concept of functional dependency. 4 *NF* is based on the concept of multivalued dependency. We can say that whenever 2 independent 1 : *M* relationships say *X* : *Y* and *X* : *Z* are there in a single relation then a multivalued dependency exists and relation is not in 4 *NF*.

Let us consider an example relation.

EMPLOYEE (**ename, projname, pnumber**)

A tuple in the above relation represents an employee whose name is ename works on a project projname has phone pnumber.

On employee can have several phones and can work on several projects. So 2 independent 1 : *M* relationship exists in employee.

ename : projname
ename : pnumber

or we can say

ename $\twoheadrightarrow$ projname
ename $\twoheadrightarrow$ pnumber

i.e. name multidetermines projname as well as pnumber. So the relation is not in 4 *NF*.

	X	Y	Z
	ename	*projname*	*pnumber*
t_1	Rahul	Insurance	266794
t_2	Rahul	Health	278977
t_3	Rahul	Insurance	278977
t_4	Rahul	Health	266794

Fig. 6.6. NOT in 4 NF

Formally, we can define *MVD* as follows *A*, *MVD* $X \rightarrow Y$, species on a given relation scheme *R*, specifies the following constraint on any relational state *r*(*R*) : if any 2 tuples t_1 and t_2 exists in *r*(*R*) such that $t_1[X] = t_2[X]$ then there should also exist 2 more tuples t_3 and t_4 with following properties.

$$t_3[X] = t_4[X] = t_1[X] = t_2[X]$$
$$t_3[Y] = t_1[Y], \; t_4[Y] = t_2[Y]$$
$$t_3[Z] = t_2[Z], \; t_4[Z] = t_1[Z]$$

where $X, Y \subseteq R$ and $Z = R - (X \cup Y)$

For the above example

$$t_3[\text{ename}] = t_4[\text{ename}] = t_1[\text{ename}] = t_2[\text{ename}] = \text{Rahul}$$
$$t_3[\text{projname}] = t_1[\text{projname}] = \text{Insurance}$$
$$t_4[\text{projname}] = t_2[\text{projname}] = \text{Health}$$
$$t_3[\text{pnumber}] = t_2[\text{pnumber}] = 278977$$
$$t_4[\text{pnumber}] = t_1[\text{pnumber}] = 266794$$

Thus, ename $\twoheadrightarrow$ projname

ename $\twoheadrightarrow$ pnumber

So we decompose the original Employee into 2 schemes

Empcon (**ename, pnumber**)

Emppro (**ename, projname**)

Both the relations are in 4 *NF*.

An *MVD* of *R* is said to be trivial *MVD* if

(*a*) $Y \subseteq X$

or (*b*) $X \cup Y = R$

Thus we can formaly define 4 *NF* as follows a relation scheme *R* is in 4 *NF* with respect to a set of dependencies *F* if, for every non-trivial *MVD* $X \rightarrow Y$ in F^+, *X* is a super key of *R*.

Or

A relation is in 4 *NF* if it is in 3 *NF* and it has no multiple sets of multivalued dependencies.

EMPLON

ename	**Dnumber**
Rahul	266794
Rahul	278977

EMPPRO

ename	*Propname*
Rahul	Insurance
Rahul	Health

Fig. 6.7. 4 NF

6.3.6. 5 NF (Fifth Normal Form)

5 NF is based on the concept of Join dependency, denoted by $JD(R_1, R_2, \ldots, R_n)$, the constraint on the relation state $r(R)$ that every legal state $r(R)$ should have a lossless join decomposition into $R_1, R_2, \ldots, R_n$.

i.e. NATURAL JOIN $\left(\pi_{R_1}(r), \pi_{R_2}(r), \ldots, \pi_{R_n}(r)\right) = r$

A relation scheme *R* is in 5 *NF* with respect to a set of functional dependency, multivalued dependency and join dependency if, for every non-trivial $JD(R_1, R_2, \ldots, R_n)$ in F^+, every *Ri* is a super key of *R*.

MVD is a special case of *JD* when $n = 2$. Let us consider an example relation.

Supply (**companyname, prodname, suppliername**)

Company supply several products to different suppliers.

Companyname	*Prodname*	*Suppliername*
Hindustan Uniliver	Paste	*X*
Hindustan Uniliver	Paste	*Y*
Hindusan Uniliver	Soap	*Y*
P & *G*	Soap	*X*
P & *G*	Shampoo	*X*
P & *G*	Shampoo	*Y*
P & *G*	Soap	*Z*

Fig. 6.8. NOT in 4 NF

We decompose it into 3 schemes.

R_1 **(companyname, prodname)**

R_2 **(companyname, suppliername)**

R_3 **(prodname, suppliername)**

R_1

Companyname	*Prodname*
Hindustan Uniliver	Paste
Hindustan Uniliver	Soap
P & *G*	Soap
P & *G*	Shampoo

R_1

Companyname	*Suppliername*
Hindustan Uniliver	*X*
Hindustan Uniliver	*Y*
P & *G*	*X*
P & *G*	Y
P & *G*	*Z*

R_3

Prodname	*Suppliername*
Paste	*X*
Paste	*Y*
Soap	*Y*
Soap	*X*
Soap	*Z*
Shampoo	*X*
Shampoo	*Y*

Fig. 6.9. NOT in 4 NF

6.4. DECOMPOSITION

As we know that normalization decompose a given relation scheme *R* (which is bad design) into more than one relation schemes. Thus decomposition is the process of replacing *R* by two or more relation schemes say R_1, R_2, ..., R_n such that these decomposed schemes are in certain normal forms. When we talk about decomposition it must satisfies the following properties.

6.4.1. Attribute Preservation Property

Attribute preservation property says that the decomposed schemas consist of subsets of attributes of original scheme *R* and all together includes all the attribute of *R*.

i.e. Each attribute in *R* will appear in at least one decomposed relation scheme.

6.4.2. Lossless Join or Non Addition Join Property

This property says that after decomposition there should be no loss of information as well as no spurious tuples are generated if we natural join the decomposed relations.

i.e. If a relation scheme R is decomposed into relations R_1, R_2, ..., R_n then for every legal state $r(R)$ under set of functional dependencies F.

$$\text{NATURAL JOIN }\left(\pi_{R_1}(r), \pi_{R_2}(r), \ldots, \pi_{Rn}(r)\right) = r$$

For binary decomposition,

$$\pi_X(r) \bowtie \pi_Y(r) = r$$

where R is decomposed in R_1 having X set of attributes and R_2 having Y set of attribute of R.

Example 6.11. Let us consider an example relation

Company (**cname, prodname, suppliername**)

cname	*prodname*	*suppliername*
P & G	P_1	S_1
H.U.	P_2	S_2
Glaxo	P_1	S_3

Fig. 6.10

If we decompose the original relation is decomposed into

R_1 (**cname, prodname**) R_2 (**prodname, suppliername**)

R_1

cname	*prodname*
P & G	P_1
H.U.	P_2
Glaxo	P_1

R_2

prodname	*suppliername*
P_1	S_1
P_2	S_2
P_1	S_3

Fig. 6.11. NOT in 4 NF

Now if take the natural join of R_1 and R_2 we get

cname	*prodname*	*suppliername*
P & G	P_1	S_1
P & G	P_1	S_3
HU	P_2	S_2
Glaxo	P_1	S_1
Glaxo	P_1	S_3

Fig. 6.12

Thus $r(R) \neq$ NATURAL JOIN (R_1, R_2).

Therefore the above decomposition is not lossless join decomposition.

A decomposition $D(R_1, R_2)$ of R is said to be lossless join decomposition with respect to a set of functional dependencies F if either

FD $\qquad R_1 \cup R_2 \longrightarrow R_1$

or *FD* $R_1 \cap R_2 \longrightarrow R_2$

i.e. attributes common to both R_1 and R_2 must contain the key for either R_1 or R_2.

6.4.3. Dependency Preservation Property

If the decomposition is not depending preserving *i.e.* if one of the dependency before decomposition gets lost after decomposition then may be some constraint cannot be applied with original relation.

Thus the decomposition must be dependency preserving.

Let us consider an example relation

CONTRACT (C#, supplier#, proj#, dep#, part#, qty, value)

Functional dependencies is given

$$F = \{C\# \rightarrow \{\text{supplier\#, proj\#, dep\#, part\#, qty, value}\}$$
$$\{\text{proj\#, part\#}\} \rightarrow C\#$$
$$\{\text{supplier\#, dep\#}\} \rightarrow \text{part\#}\}$$

If we decompose CONTRACT into

R_1 (C#, supplier#, proj#, dep#, qty, value)

R_2 (**supplier#, dep#,** past#)

Then the decomposition is not dependency pressing because neither R_1 or R_2 have the dependency {proj#, part#} → *C#*. Thus, the decomposition of a relation schema *R*.

Into R_1, R_2, ..., R_m with respect to set of functional dependencies *F* is dependency preserving if

$$\left(\pi_{R_1}(F) \cup \pi_{R_2}(F) \cup .... \cup (\pi_{R_m}(F)\right)^+ = F^+$$

where $\pi_{R_1}(F)$ is the projection of *F* on R_i.

A decomposition of relation scheme *R* into R_1 and R_2 with respect to attribute set *X* and *Y* respectively w.r.t. set *F* is dependency preserving if

$$(F_X \cup F_Y)^+ = F^+$$

where F_X = projection of *F* on *X*

F_Y = projection of *F* on *Y*

Projection of *F* on *X* is the set of functional dependencies in F^+ that involves only attributes in set *X*. *i.e.* $U \rightarrow V$ in F^+ is in F_X only if all the attributes in *U* and *V* are in *X*.

Example 6.12. Let us consider an example relation

$$R\ (ABC) \text{ with } F = \{A \rightarrow B,\ B \rightarrow C,\ C \rightarrow A\}$$

If we decompose *R* into R_1 (*AB*) R_2 (*BC*) then find whether it is depending preserving or not.

Solution : Given *R* (*ABC*)

R_1 (*AB*) R_2 (*BC*)

were *X* = *AB* *Y* = *BC*

First of all find F^+

We get

$$F^+ = \{A \rightarrow B,\ B \rightarrow C,\ C \rightarrow A,$$
$$B \rightarrow A \qquad (\because B \rightarrow C \text{ and } C \rightarrow A)$$
$$C \rightarrow B \qquad (\because C \rightarrow A \text{ and } A \rightarrow B)$$
$$A \rightarrow C \qquad (\because A \rightarrow B \text{ and } B \rightarrow C)$$

Now compte $(F_X \cup F_Y)^+$

Note F_X is the projection of F on X

Here $X = AB$

So, $F_{AB} = \{A \to B, B \to A\}$

Similarly, $F_{BC} = \{B \to C, C \to B\}$

Now, $F_{AB} \cup F_{BC} = \{A \to B, B \to A, B \to C, C \to B\{$

Again $(F_{AB} \cup F_{BC})^+ = \{A \to B, B \to A, B \to C, C \to B, A \to C, C \to A\}$

Thus $(F_{AB} \cup F_{BC})^+ = F^+$

Hence it is dependency preserving.

Summary

- Normalization is the process to refinement of the data base design.
- Normalization takes the bad database design (the design which suffers from data redundancy, insertion anomalies, deletion and updation anomalies is called as bad design) and decomposed it into one of the normal form.
- Certain tests are there for various normal forms.
- Normalization is carried out with respect to functional dependency, multivalued dependency join dependency.
- In normalization, we take a relation scheme R and decompose its into 2 or more schemes that are in certain normal form.
- Normalization is a technique that is used to design relational databases, basically, with Normalization we design a logical structure of the database. Normalization is based on the concepts of normal forms.
- Thus normalization is a process of decompoing the redundant relation schemas by breaking up their attributes into smaller relation schemas that poses deriable properties.
- In this process the database design that (which is bad form) suffers from data redundancy, various anomalies is decomposed into one or more relation schemes, i.e., we take a relation scheme R and decompose it into 2 or more relation schemes say R_1, R_2 ... etc.

Multiple Choice Questions

1. In normalization, the process of decomposition is called as

(a) Normalization (b) Gneralization

(c) Aggregation (d) Specification

2. An association between 2 set of attributes is called as

(a) Join dependency (b) Association

(c) Functional dependency (d) Multivalued dependency

3. In INF, the value of the attribute must be

(a) Atomic (b) Simple

(c) Domestic (d) None

4. A table which is 3 NF, must be in

(a) 1 NF (b) 2 NF

(c) BCNF (d) None

5. The process of Normalization is

(*a*) Iterative (*b*) Non-reversible

(*c*) Reversible (*d*) None

6. Which normal form is most desirable?

(*a*) DKNF (*b*) 3 NF

(*c*) BCNF (*d*) 4 NF

7. A tuple is a

(*a*) Column of a table (*b*) Two dimensional table

(*c*) rowx of the table (*d*) Key

8. The highest normal form a relation is in if all its non-key attributes are dependent on all of the key :

(*a*) 1 NF (*b*) 2 NF

(*c*) 3 NF (*d*) 4 NF

9. A functional dependency is a relationship between

(*a*) tables (*b*) columns

(*c*) attributes (*d*) rows

10. Which of the following is a group of one or more attributes that uniquely identifies a row ?

(*a*) key (*b*) tuple

(*c*) determinant (*d*) relation

11. A relation in this form is free of all modification anomalies

(*a*) 1 NF (*b*) 2 NF

(*c*) 3 NF (*d*) DKNF

12. Spurious tupe may occur due to

(*a*) theta join (*b*) Bad normalization

(*c*) updating tables join (*d*) (*a*) and (*b*)

(*e*) (*a*) and (*c*) (*f*) (*a*) and (*c*)

(*g*) (*b*) and (*c*)

13. Two or more attributes that can be a key are called :

(*a*) determinants (*b*) primary key

(*c*) composite key (*d*) foreign key

14. A relation is in this form if it is in BCNF and has no multivalued dependency

(*a*) 2 NF (*b*) 4 NF

(*c*) 3 NF (*d*) DKNF

15. A relation schema *R* is in .... if it is in 3 NF and satisfies an additional constant that for every *FD* $X \rightarrow A$. X must be a candidate key.

(*a*) 1 NF (*b*) 2 NF

(*c*) 3 NF (*d*) BCNF

Selected Exercises

1. What is the difference between 3 NF and BCNF ?

2. What is the need of normalization ?

3. Whether data redundancy can be eliminated completely ? If not, why ?

4. What is denormalization ?
5. Write down the Armstrong's rule.
6. What undesirable dependencies are avoided when relation its in
   (*i*) 2 NF
   (*ii*) 3 NF.
7. What do you mean by functional dependency? differentiate between full functional dependency and partial functional dependency?
8. What are the characteristics of functional dependency?
9. What do you mean by trivial and non trivial functional dependency? Give example of each.
10. Given FDS as,
    $A \to BC, \quad CD \to E, \quad B \to D, \quad B \to A$, compute B^+?
11. Given the set F of FDS for R (A, B, C, D, E)
    $F = \{A \to BC, \quad CD \to E, \quad B \to D, \quad E \to A\}$
    (*i*) Copute the closure of F, F^+
    (*ii*) Canonical cover, F_C
12. Given 2 sets F_1 and F_2 of FDS for a relation
    $F_1 : A \to B, \quad AB \to C, \quad D \to AC, \quad D \to E$
    $F_2 : A \to BC, A \to AE,$
    Are these 2 sets equivalent?
13. Given F: $A \to BC, E \to CF, \quad B \to E, \quad C \to EF$ compute $\{AB\}^+$?
14. Given R (A, B, C, D, E) with dependencies F: $AB \to C, CD \to E, \quad DE \to B$
    (*i*) is AB a candidate key?
    (*ii*) is ABD a candidate key?
15. Given $R(A, B, C, D)$ has FDS
    $F : AB \to C, D \to A, \quad AE \to B, \quad CD \to E, \quad BE \to D$ find all the candidate keys of R?
16. Consider the relation
    R . $(A, B, C, D, E, F, G, H, I, J)$ with
    $F : AB \to C, \quad A \to DE, \quad B \to F, \quad F \to GH, \quad D \to IJ$
    (*i*) What is the key for R?
    (*ii*) Decomposes R into 2 NF.
17. What are the advantages of normalized relations over unnormalized relation? Explain with an example?
18. What is decomposition? Explain lossless and lossy decomposition with examples.
19. With on example differentiate 3NF and BCNF.
20. Is BCNF stronger than 3 NF? Justify your answer.
21. Consider a relation R (A, B, C, D, E) with FDS
    $A \to BC, \quad C \to D, \quad D \to B, \quad B \to E, \quad A \to E$
    (*i*) Give loss less join dcomposition of R
    (*ii*) Whether decomposition is dependency preserving or not?
22. Given relation R (A, B, C, D, E, F, G, H) with FDS
    $F : \quad A \to BCD, \quad AE \to F, \quad E \to G, \quad D \to H$
    (*i*) Find the key of R.
    (*ii*) In which the form R is?
    (*iii*) Decompose R until it can not be decomposed further.

23. Consider R $(A, B, C, D, E, F, G, H, I)$ with FDS

$F' : AB \rightarrow C, \quad A \rightarrow DE, \quad B \rightarrow F, \quad F \rightarrow GH, \quad D \rightarrow IJ$

(*i*) What is the key of R?

(*ii*) Decompose R into 2NF?

(*iii*) Decompose R into 3NF?

24. Prove that any relation schema with two attributes is in BCNF always.

25. Wht is the functional dependency ? List all functional dependencies satisfied by relation given below :

R

A	B	C
a_1	b_1	c_1
a_1	b_1	c_2
a_2	b_1	c_1
a_1	b_1	c_3

26. Consider the scheme R $(ABCDE)$ with following *FDS*

$$\{E \rightarrow A, \; CD \rightarrow E, \; A \rightarrow BC, \; B \rightarrow D\}$$

state whether the following decomposition are loss less join decompositions or not

(*i*) $\{(ABC) \; (ADE)\}$

(*ii*) $\{(ABC) \; (CDE)\}$

27. What is the motivation behind normalizing a database ?

28. Discuss join dependency and 5NF. Why is 5NF also called pJNF ?

29. "Redundancy of data is many time beneficial". Justifly the statement; also describe the situation when redundancy will nuss up the current data base step.

30. Why 4NF is more desriable than BCNF.

❑❑❑

Chapter

7 Oracle 8 Architecture

Inside this Chapter

7.0. ORACLE 8 ARCHITECTURE

Oracle database is a collection of data treated as a single unit is oracle database may be client/server or multitier. **When client/server architecture** used the database system is divided into 2 part a front end (client) and back end (server).

The client initiates a request that is to be handled by the server, client is running user interface programs and application programs.

Whereas the serves run the oracle software and handles all incoming concurrent data requests.

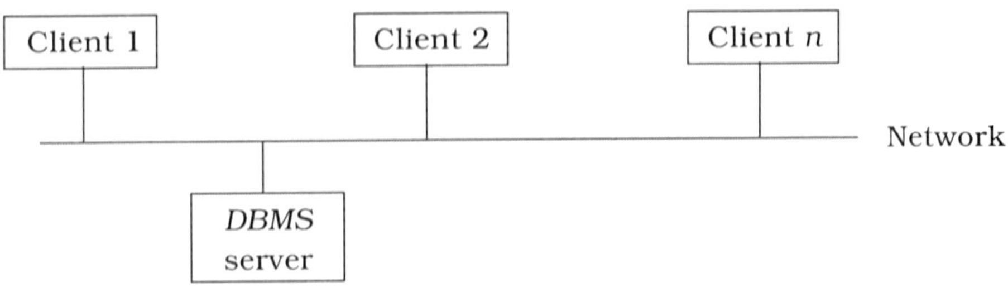

Client/server **Architecture**

When multitier (*i.e.* three ties for example) architecture to used an intermediate middle tier (application serves or web server) is there in between client and the server.

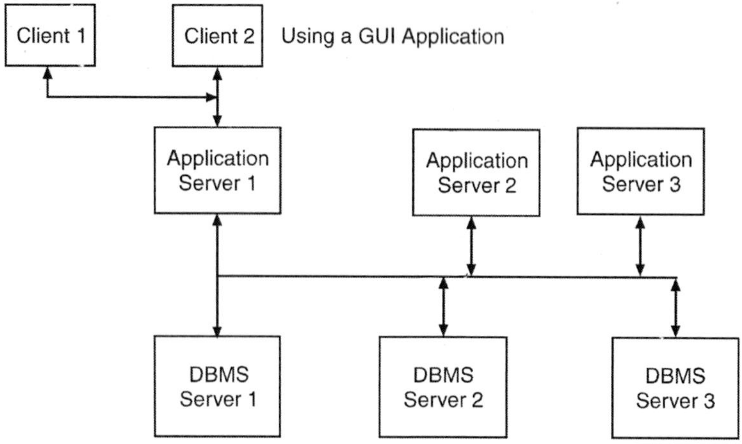

Clients start operations by having a connection with one or more application servers. Application server implements the Business Logic. Application server validate the client, contect to oracle data base server and perform the requested operation on behalf of the client.

Oracle 8 INSTANCE Architecture :

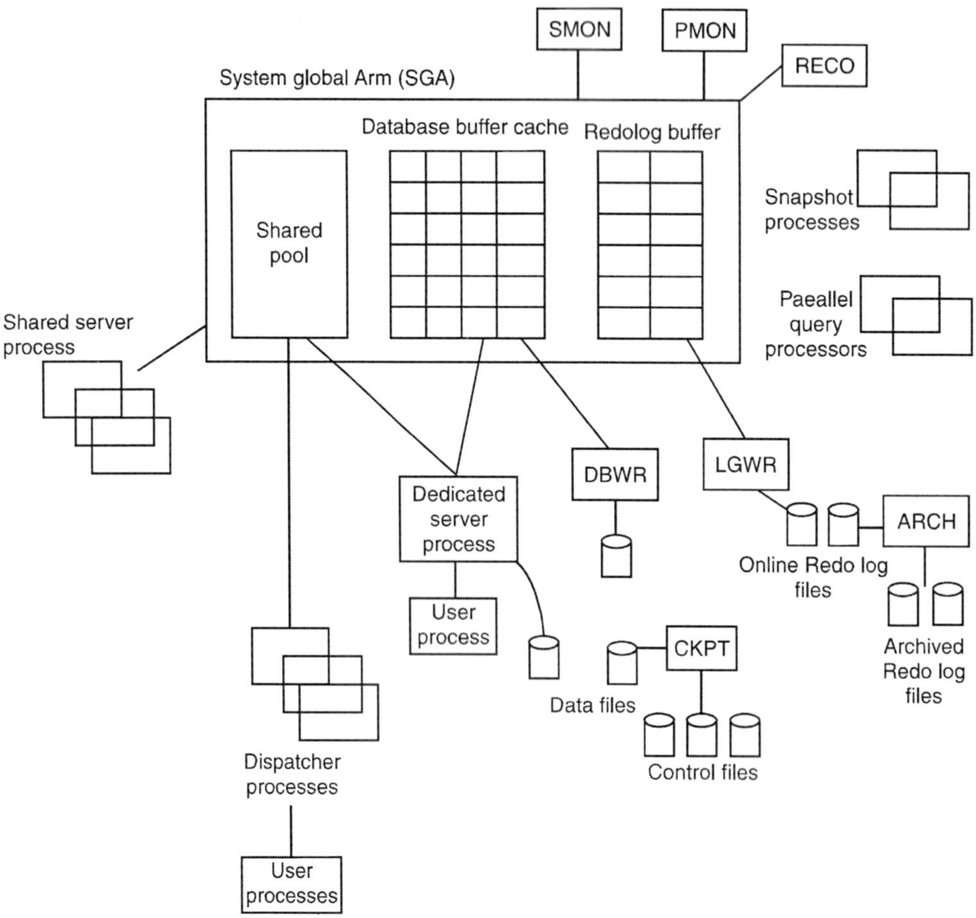

7.1. SYSTEM ARCHITECTURE

In the following sections we discuss the main components of the oracle DBMS (Version 7.X) architecture (Section 4.6.1) and the logical and physical database structures (Sections 4.6.2 and 4.6.3). We furthermore sketch how SQL statements are processed (Section 4.6.4) and how database objects are created (Section 4.6.5).

7.1.1. Storage Management and Processes

The ORACLE DBMS server is based on a so-called **Multi-Server Architecture**. The server is responsible for processing all database activities such as the execution of SQL statements, user and resource management, and storage management. Although there is only one copy of the program code for the DBMS server, to each user connected to the server logically a separate server is assigned. The following figure illustrates the architecture of the oracle DBMS consisting of storage structures, processes, and files.

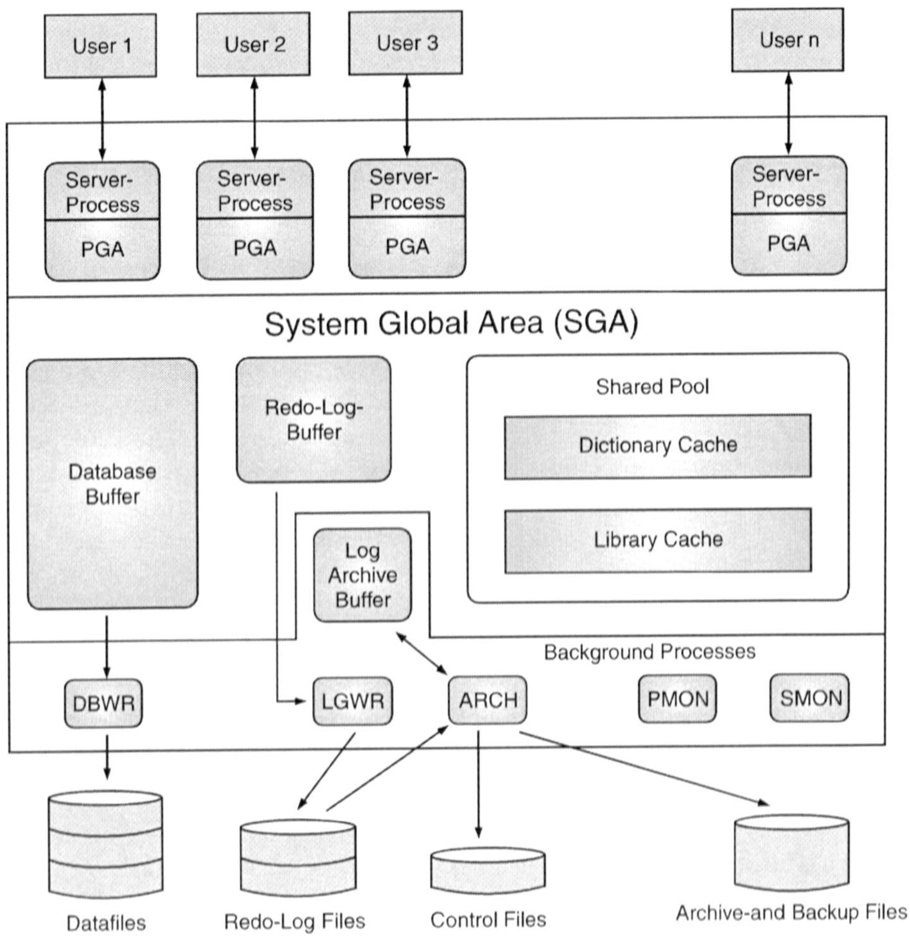

Fig. 7.1. Oracle system architecture

Each time a database is started on the server (*instance startup*), a portion of the computer's main memory is allocated, the so-called **System Global Area (SGA)**. The SGA consists of the **shared pool**, the **database buffer**, and the **redo-log buffer**. Furthermore, several background processes are started. The combination of SGA and processes is called **database instance**. The memory and processes associated with an instance are responsible for efficiently managing the data stored in the database, and to allow users accessing the database concurrently. The oracle server can manage multiple instances, typically each instance is associated with a particular application domain.

The SGA serves as that part of the memory where all database operations occur. If several users connect to an instance at the same time, they all share the SGA. The information stored in the SGA can be subdivided into the following three caches.

Database Buffer: The database buffer is a cache in the SGA used to hold the data blocks that are read from data files. Blocks can contain table data, index data etc. Data blocks are modified in the database buffer. oracle manages the space available in the database buffer by using a least recently used (LRU) algorithm. When free space is needed in the buffer, the least recently used blocks will be written out to the data files. The size of the database buffer has a major impact on the overall performance of a database.

Redo-Log-Buffer: This buffer contains information about changes of data blocks in the database buffer. While the redo-log-buffer is filled during data modifications, the log writer process writes information about the modifications to the redo-log files. These files are used after, e.g., a system crash, in order to restore the database (database recovery).

Shared Pool: The shared pool is the part of the SGA that is used by all users. The main components of this pool are the dictionary cache and the library cache. Information about database objects is stored in the data dictionary tables. When information is needed by the database, for example, to check whether a table column specified in a query exists, the dictionary tables are read and the data returned is stored in the dictionary cache. Note that all SQL statements require accessing the data dictionary. Thus keeping relevant portions of the dictionary in the cache may increase the performance. The library cache contains information about the most recently issued SQL commands such as the parse tree and query execution plan. If the same SQL statement is issued several times, it need not be parsed again and all information about executing the statement can be retrieved from the library cache.

Further storage structures in the computer's main memory are the **log-archive buffer** (optional) and the **Program Global Area** (PGA). The log-archive buffer is used to temporarily cache redo-log entries that are to be archived in special files. The PGA is the area in the memory that is used by a single oracle user process. It contains the user's context area (cursors, variables etc.), as well as process information. The memory in the PGA is not sharable.

For each database instance, there is a set of processes. These processes maintain and enforce the relationships between the database's physical structures and memory structures. The number of processes varies depending on the instance configuration. One can distinguish between user processes and oracle processes. ORACLE processes are typically background processes that perform I/O operations at database run-time.

DBWR: This process is responsible for managing the contents of the database buffer and the dictionary cache. For this, DBWR writes modified data blocks to the data files. The process only writes blocks to the files if more blocks are going to be read into the buffer than free blocks exist.

LGWR: This process manages writing the contents of the redo-log-buffer to the redo-log files.

SMON: When a database instance is started, the system monitor process performs instance recovery as needed (e.g., after a system crash). It cleans up the database from aborted transactions and objects involved. In particular, this process is responsible for coalescing contiguous free extents to larger extents (space defragmentation, see Section 4.6.2).

PMON: The process monitor process cleans up behind failed user processes and it also cleans up the resources used by these processes. Like SMON, PMON wakes up periodically to check whether it is needed.

ARCH: (optional): The LGWR background process writes to the redo-log files in a cyclic fashion. Once the last redo-log file is filled, LGWR overwrites the contents of the first redo-log file. It is possible to run a database instance in the archive-log mode. In this case the ARCH process copies redo-log entries to archive files before the entries are overwritten by LGWR. Thus it is possible to restore the contents of the database to any time after the archive-log mode was started.

USER: The task of this process is to communicate with other processes started by application programs such as SQL*Plus. The USER process then is responsible for sending respective operations and requests to the SGA or PGA. This includes, for example, reading data blocks.

7.1.2. Logical Database Structures

For the architecture of an oracle database we distinguish between logical and physical database structures that make up a database. Logical structures describe logical areas of storage (name spaces) where objects such as tables can be stored. Physical structures; in contrast, are determined by the operating system files that constitute the database.

The logical database structures include:

Database: A database consists of one or more storage divisions, so-called tablespaces.

Tablespaces: A tablespace is a logical division of a database. All database objects are logically stored in tablespaces. Each database has at least one tablespace, the SYSTEM tablespace, that contains the data dictionary. Other tablespaces can be created and used for different applications or tasks.

Segments: If a database object (e.g., a table or a cluster) is created, automatically a portion of the tablespace is allocated. This portion is called a segment. For each table there is a table segment. For indexes so-called index segments are allocated. The segment associated with a database object belongs to exactly one tablespace.

Extent: An extent is the smallest logical storage unit that can be allocated for a database object, and it consists a contiguous sequence of data blocks! If the size of a database object increases (e.g., due to insertions of tuples into a table), an additional extent is allocated for the object. Information about the extents allocated for database objects can be found in the data dictionary view USER_EXTENTS.

A special type of segments are **rollback segments**. They don't contain a database object, but contain a "before image" of modified data for which the modifying transaction has not yet been committed. Modifications are undone using rollback segments. oracle uses rollback segments in order to maintain read consistency among multiple users. Furthermore, rollback segments are used to restore the "before image" of modified tuples in the event of a rollback of the modifying transaction.

Typically, an extra tablespace (RBS) is used to store rollback segments. This tablespace can be defined during the creation of a database. The size of this tablespace and its segments depends on the type and size of transactions that are typically performed by application programs.

A database typically consists of a SYSTEM tablespace containing the data dictionary and further internal tables, procedures etc., and a tablespace for rollback segments. Additional tablespaces include a tablespace for user data (USERS), a tablespace for temporary query results and tables (TEMP), and a tablespace used by applications such as SQL*Forms (TOOLS).

7.1.3. Physical Database Structures

The physical database structure of an oracle database is determined by files and data blocks:

Data Files: A tablespace consists of one or more operating system files that are stored on disk. Thus a database essentially is a collection of data files that can be stored on different storage devices (magnetic tape, optical disks etc.). Typically, only magnetic disks are used. Multiple data files for a tablespace allows the server to distribute a database object over multiple disks (depending on the size of the object).

Blocks: An extent consists of one or more contiguous oracle data blocks. A block determines the finest level of granularity of where data can be stored. One data block corresponds to a specific number of bytes of physical database space on disk. A data block size is specified for each oracle database when the database is created. A database uses and allocates free database space in oracle data blocks. Information about data blocks can be retrieved from the data dictionary views USER_SEGMENTS and USER_EXTENTS. These views show how many blocks are allocated for a database object and how many blocks are available (free) in a segment/extent.

As mentioned in Section 4.6.1, aside from datafiles three further types of files are associated with a database instance:

Redo-Log Files: Each database instance maintains a set of redo-log files. These files are used to record logs of all transactions. The logs are used to recover the database's transactions in their proper order in the event of a database crash (the recovering operations are called roll forward). When a transaction is executed, modifications are entered in the redo-log buffer, while the blocks affected by the transactions are not immediately written back to disk, thus allowing optimizing the performance through batch writes.

Control Files: Each database instance has at least one control file. In this file the name of the database instance and the locations (disks) of the data files and redo-log files are recorded. Each time an instance is started, the data and redo-log files are determined by using the control file(s).

Archive/Backup Files: If an instance is running in the archive-log mode, the ARCH process archives the modifications of the redo-log files in extra archive or backup files. In contrast to redo-log files, these files are typically not overwritten.

The following ER schema illustrates the architecture of an oracle database instance and the relationships between physical and logical database structures (relationships can be read as "consists of").

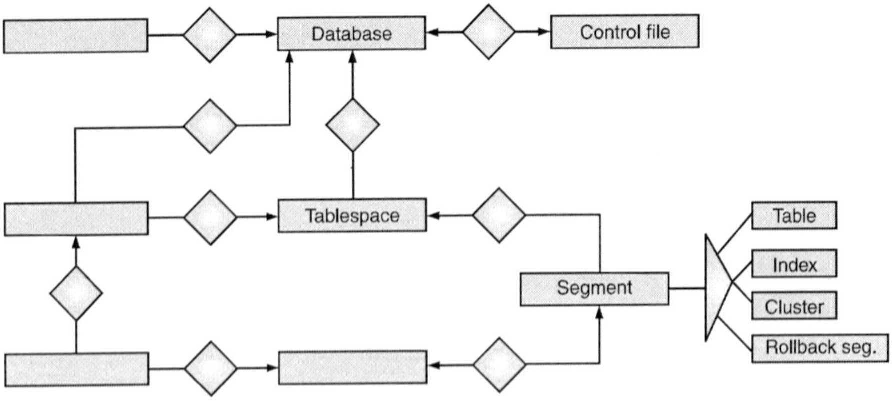

Fig. 7.2. Relationships between logical and physical database structures.

7.1.4. Steps in Processing an SQL Statement

In the following we sketch how an SQL statement is processed by the oracle server and which processes and buffers involved.

1. Assume a user (working with SQL*Plus) issues an update statement on the table TAB such that more than one tuple is affected by the update. The statement is passed to the server by the USER process. Then the server (or rather the query processor) checks whether this statement is already contained in the library cache such that the corresponding information (parse tree, execution plan) can be used. If the statement cannot be found, it is parsed and after verifying the statement (user privileges, affected tables and columns) using data from the dictionary cache, a query execution plan is generated by the query optimizer. Together with the parse tree, this plan is stored in the library cache.
2. For the objects affected by the statement (here the table TAB) it is checked, whether the corresponding data blocks already exist in the database buffer. If not, the USER process reads the data blocks into the database buffer. If there is not enough space in the buffer, the least recently used blocks of other objects are written back to the disk by the DBWR process.
3. The modifications of the tuples affected by the update occurs in the database buffer. Before the data blocks are modified, the "before image" of the tuples is written to the rollback segments by the DBWR process.
4. While the redo-log buffer is filled during the data block modifications, the LGWR process writes entries from the redo-log buffer to the redo-log files.

5. After all tuples (or rather the corresponding data blocks) have been modified in the database buffer, the modifications can be committed by the user using the **commit** command.
6. As long as no **commit** has been issued by the user, modifications can be undone using the **rollback** statement. In this case, the modified data blocks in the database buffer are overwritten by the original blocks stored in the rollback segments.
7. If the user issues a **commit,** the space allocated for the blocks in the rollback segments is deallocated and can be used by other transactions. Furthermore, the modified blocks in the database buffer are unlocked such that other users now can read the modified blocks. The end of the transaction (more precisely the **commit)** is recorded in the redo-log files. The modified blocks are only written to the disk by the DBWR process-if the space allocated for the blocks is needed for other blocks.

7.1.5. Creating Database Objects

For database objects (tables, indexes, clusters) that require their own storage area, a segment in a tablespace is allocated. Since the system typically does not know what the size of the database object will be, some default storage parameters are used. The user, however, has the possibility to explicitly specify the storage parameters using a storage clause in, e.g., the create table statement. This specification then overwrites the system parameters and allows the user to specify the (expected) storage size of the object in terms of extents.

Suppose the following table definition that includes a storage clause:

```
create table STOCKS
    (ITEM        varchar2(30),
    QUANTITY number(4))
storage (initial 1M next 400k
        minextents 1 maxextents 20 pctincrease 50);
```

initial and **next** specify the size of the first and next extents, respectively. In the definition above, the initial extent has a size of 1 MB, and the next extent has a size of 400 KB. The parameter **minextents** specifies the total number of extents allocated when the segment is created. This parameter allows the user to allocate a large amount of space when an object is created, even if the space available is not contiguous. The default and minimum value is 1. The parameter **maxextents** specifies the admissible number of extents. The parameter **pctincrease** specifies the percent by which each extent after the second grows over the previous extent. The default value is 50, meaning that each subsequent extent is 50% larger than the preceding extent. Based on the above table definition, we thus would get the following logical database structure for the table STOCKS (assuming that four extents have already been allocated):

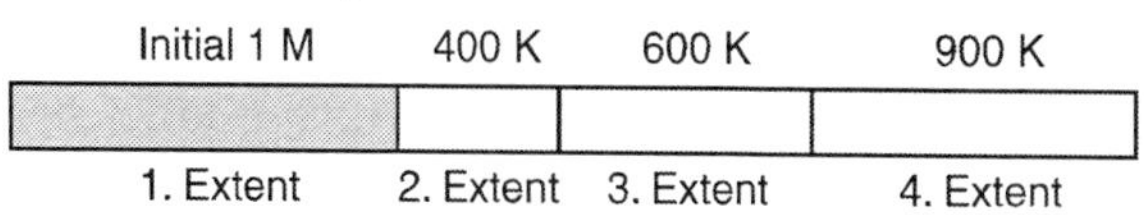

Fig. 7.3. Logical Storage Structure of the Table STOCKS.

If the space required for a database object is known before creation, already the initial extent should be big enough to hold the database object. In this case, the oracle server (more precisely the resource manager) tries to allocate contiguous data blocks on disks for this object, thus the defragmentation of data blocks associated with a database object can be prevented.

For indexes a storage clause can be specified as well

```
create index STOCK_IDX on STOCKS (ITEM)
        storage (initial 200k next 100k
                minextents 1 maxextents 5).
```

7.2. DATABASE STORAGE

We define the oracle database as the configuration files, data files, control files and redo log files as well as the table indexes etc.

The database can be categorised into 2 categories.

7.2.1. System Database Objects

System database objects are the objects that are used internally by *RDBMS*. They are used for supporting internal database functions and thus are created and configured either by *DBA* or by server itself. They are

initialization parameter file(s)

Control file

Online redo log files

Archieved redo logs

The trace file(s)

The ROWID

Oracle blocks

7.2.1.1. Initialization Parameter File

Initialization parameter file or init. ora is the primary configuration file and used to contain configuration keys and values each of which control the operation of a database By default having a name init, D.Ora and stored in SORACLE_HOME/database directory on WINNT serves.

7.2.1.2. Control Files

Contain information on what data files and redo log files belong to the database, what character set data should be stored etc.

Control file is created when the database is created. Most of the parameters contained in the control file are set during the database creation.

Some of the modifiable parameters are

MAX DATA FILES → maximum number of data files

MAX LOG FILES → maximum number of online redo log files

7.2.1.3. Online Redo Log Files

The *LGWR* (log wirter background process) with the contents of the redo log cache to these online redo log files. The redo log file store all the changed information for the database and are used for recovery mechanism.

7.2.1.4. Trace File(s)

Oracle database have at least one file where system messages errors and major events are logged. This sid ALRT . log is stored in at the location specified by init. ora.

Background and useuser processes create their own trace files whenever problems or failures occur.

7.2.1.5. ROWID

To retrieve information from oracle database it must be uniquely Identifiable in the database. ROWID do this identification. ROWID is the internal structure and is a 2 byte value that stores the physical location for a row in the database.

The format is

BBBBBBB . RRRR . FFFF

BBBBBBB	RRRR	FFFF
tells block number where rows resides in the data file	row number in the block where the data row exists	the file where the block exist

Whenever a row is created it is assigned a ROWID and it remains the same until the row is deleted.

7.2.1.6. Oracle Blocks

Oracle blocks is the smallest unit of storage the server can address. All data occurs is performed in terms of oracle block. Each block contains a header information future update of data, and the actual rows stored.

7.2.2. Database Objects

Database objects are objects that can be accessed by any process. It includes data files, extents, table space and database segments.

7.2.2.1. Data Files

Data files are the only physical database objects. Each data file is allocated to one table space and holds the actual data stored in that table space. The data stored in the data file is in oracle binary format.

7.2.2.2. Extents

An extent is a collection of one or more contiguous oracle blocks and databae segments is composed of one or more extents.

A database segment is allocated on extent upon object creation. If a segment no longer fit new data into its currently allocated extent, it must allocate another extent.

7.2.2.3. Table Splace

A table space is logical structure used to group similarly accessed data. Each table space is made up of one or more data files all objects must specify a table space whenever they are created.

7.2.2.4. Database Segments

Database segments are the user created objects that are stored in the database. That includes, tables, indexes, table cluster so on.

7.3. ORACLE SOFTWARE STRUCTURES

When we install the oracle software, on oracle directory structure is created to store the executables, shared libraries, configuration files, trace files and so on both server and client software is stored in the same directory.

Oracle is installed on the server using on operating system account created specially for the task oracle has a directory structure.

7.3.1. Oracle Home

Oracle home is the top level directory at oracle directory tree. It is defined as environment variable or registry entry name ORACLE_HOME. Many executable use this variable to locate shared code, configuration files etc.

This ORACLE_HOME directory contains other directories, each of which contains oracle files or other directories they are

BIN — Contain executables and shared DLLs.

DBS — Oracle serves files and default directory for init. ora files.

DATABASE — Starter database and default oracle database file.

RDBMS — Database files.

RDBMS/ADMIN — Sql. scripts to create system database objects, roles and views.

RBMS/TRACE — It is default database dump directory.

NETWORK — Contains network files.

NETWORK/ADMIN — Contains SQL* net configruation files.

NETWORK/TRACE — Default SQL* net dump directory.

PLSQL — Contains PL/SQL configuration files and SQL scripts.

DOC — Contains online documents.

PLUS — Contains SQL* plus files.

7.3.2. Other Configuration Files

The password file oraSID. pwd created by oradim utility on NT server. This file contains the password for internal user account and is flagged and hidden.

7.4. SHARED DATABASE ACCESS MECHANISM

Oracle provides oracle Net services that are used to interact with communicating protocols such as *TCP/IP, HTTP, FTP.*

Oracle net services has a component oracle Net that create and maintains a network session from a client application to an oracle database server. Oracle Net is a software layer that resides on both the client application and on the database server. Oracle net provides scalcibility by maximizing system resources and thus improve the overall system performance.

Oracle's shared database access mechanism increases the application's scalability as well as it allows database server to be connected with number of clients concurrently.

In oracle's shared database architecture, client applications do not communicate directly with database's server process rather each client request is routed to one or more dispatcher.

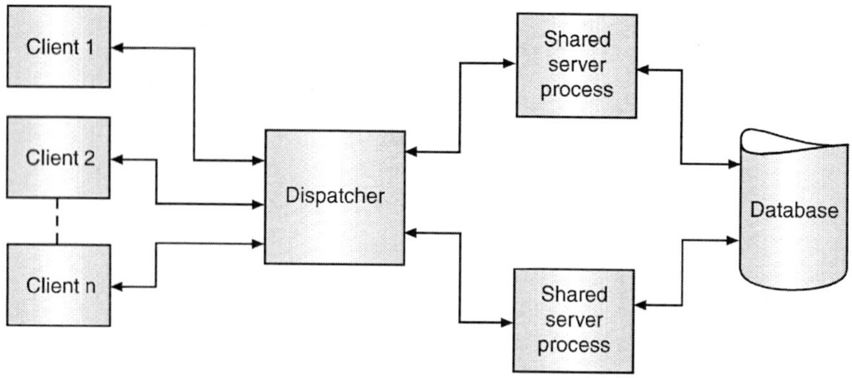

Fig. 7.4. Shared database Access Architecture

The dispatcher maintains a client request quene. An idle shared server, from the available shared pool of server processes, fetchs the request and then process the request.

7.5. DATABASE PROTECTION

The primary resource of the oracle database night sys. is the oracle database itself. Therefore database must be protected from any unauthorised access and/or accidental damage for that, oracle uses various schemes and security mechanism such as user authentication, database privilege management, monitoring the database *i.e.* database Audit, Hardware security and recovery mechanism.

7.5.1. User Authentication

Before database is allowed to access, its all users must be identified and authenticated by the oracle database. User authentication can be done by external and by enterprise.

7.5.1.1. Database Authentication

Database authentication is done when a user is created and password is specified. Oracle assign a profile to auser for password management. This profile can be used to enforce rules such as

1. if user has failed 3 logins, the account is locked.
2. given password may have a specified time limit for use and will expire there need to change.
3. complexity verification checks the strength of the password.

7.5.1.2. External Authentication

Can be obtained either by operating system or network authentication service. A database password is not required for this type of login. Set the parameter OS-AUTHENT-PREFIX in the file init-ora.

Whereas network authentication can be accomplished with *ANO* (Advanced Networking Option) and using the following technologies

— Network authentication services such as Kerberos and sesame.

— Token devices.

— Biometric devices.

7.5.1.3. Enterprise Authentication

Enterprise authentication allows a global user for password management and enforce single sign on using Oracle Security Service (*OSS*). *OSS* will interface with oracle enterprise manager to centralize security role management and enterprise authorization.

When a global user logon to a database, the global roles are dynamically assigned to that user.

7.5.2. Database Privilegs Management

There are 2 kind of privileges namely system privilege and object privilege.

System privileges allows uses to perform administrative activites. Thus to be given to administrator only. These are

ANALYSE ANY : analyze any table, cluster, or index in the database.

AUDI ANY : audit any schema object in the database.

CREATE PROFILE : create profiles.

CREATE ROLE : create roles.

There are many more system privileges.

Whereas **object privileges** allows users to access and maintain the database objects. So the are assigned to sad users they are

ALTER, DELETE, EXECUTE, INDEX, INSERT, SELECT, UPDATE.

7.5.3. Monitoring Database

Oracle can audit and record activities that are happing on the database. There are 3 categories of audit.

7.5.3.1. Auditing Login

Can audit all succssful and unsuccessful attempts to establish a connection with the database.

AUDIT SESSION WHENEVER NOT SUCCESSFUL

AUDIT SESSION WHENEVER SUCCESSFUL

Commands are used. AUDIT ANY privilege is required to execute the above command.

7.5.3.2. Audit Database Actions

You can audit at the statement and system privilege level. The statement audit options are ALTER SYSTEM, NOT EXISTS, SEQUENCE.

7.5.3.3. *You can audit a specific schema objects also*

e.g.

```
AUDIT insert
ON scott. emp
WHENEVER NOT SUCCESSFUL.
```

7.5.4. Protecting Data Integrity

Data integrity is protected by the use of secure communication channel and by applying appropriate encryption algorithms.

7.5.5. Hardware Security

All the necessary hardware required for database is stored in a restricted area where access is allowed only for limited and authorised persons. Power is backed up by having power full ups.

7.5.6. Recovering Cost Data

If a failure occur due to any security breach then recovery mechanism must handle from this failure.

Summary

- The term database to used for both the entire database management environment and a description of logical and physical data structures that make up a *RDBMS*.
- Opening on oracle database involves 3 steps
  1. Creating the oracle instance.
  2. Mounting the database by the instance.
  3. Opening the database.
  4. Default accounts are installed whenever you install oracle namely SYS and SYSTEM.
- Sys is the owner of all internal database tables, structures, package, procedures. *i.e.* He/her is the root or administrator of oracle database.
- Sys is having password CHANGE-ON-INTALL.
- SYSTEM has full rights to all database objects default password is MANAGER.
- After installation you should change password for both SYS and SYSTEM.

Multiple Choice Questions

1. Oracle is a

(*a*) NDBMS (*b*) BDBMS
(*c*) RDBMS (*d*) None

2. SGA stands for

(*a*) System global area (*b*) System Go area
(*c*) System Go ahead (*d*) None

3. SGA has

(*a*) 4 caches (*b*) 5 caches
(*c*) 3 caches (*d*) 2 caches

4. A database is divided into logical storage units called as
   (a) Tables (b) Records
   (c) Tablespaces (d) None

5. Oracle server consists of
   (a) Oracle database + Oracle instance (b) Oracle instances only
   (c) Oracle database + DBMS (d) None

Selected Exercises

1. Why background processes are there in oracle ?
2. What is the significance of oracle data dictionary ?
3. What is the significance of Oracle data dictionary?
4. Explain the architecture of RDBMS Oracle in brief?
5. Why background processes are there in Oracle?
6. What mechanisms are used for
   (a) database protection
   (b) accessing shared database
7. Explain the following
   (a) SMON
   (b) PMON
8. What do you mean by ROWID? What is its significance?

❐❐❐

Chapter

8 Transaction Processing Concepts

Inside this Chapter

8.1. TRANSACTION SYSTEM

"Collections of operations that form a single logical unit of work are called transaction."

A transaction is a unit of program execution that accesses and possibly updates various data items. It is also define as a logical unit of database processing that includes one or more database access operations.

Transactions access data using two operations :

1. **Read (X)** : Which transfers the data item (X) from the database to a local buffer belonging to the transaction that executed the read operation.
2. **Write (X) :** Which transfers the data item (X) from the local buffer of the transaction that executed the write back to the database.

8.2. PROPERTIES OF TRANSACTION

Transactions have four basic properties, which are called ACID properties. These properties are closely related to each other.

A → Atomicity
C → Consistency
I → Isolation
D → Durability

Atomicity

A transaction is an atomic unit of processing. It is either performed completely or not performed at all. At the need of a transaction the updates made by the transaction will be accessible to other transaction.

Consistency

The consistency property of a transaction implies that if database was in a consistent state before the initiation of a transaction, then at the end of transactions the database will also be in a consistent state.

Isolation

The isolation property of a transaction indicates that the steps or operations performed by a transaction should be isolated from other transaction, i.e., the execution of a transaction should not be interfered with by any other transaction being executed concurrently.

Durability

The changes applied to the database by a committed transaction must persist in the database. These changes must not be lost because of any failure.

Example : In a Banking system there are many accounts are there and a set of transactions that access and updates those accounts. For the time we assume that the database permanently resides on disk, but that some portion of it is temporarity residing in main memory.

The following two opearations are accompklished to access the database :

- Fead (X), which transfers the data item X from the database to a local buffer belonging to the transaction that execuited the **read** operation.
- Write (X), which transfer the data item X from the local buffer of the transaction that executed the **write** back to the database.

In real Database system, the write operation does not necessarily result in the immediate update of the data on disk, the write operation may be temporarily stored in memory and executed on the disk later. However we shall assume that the write operation updates the database immediately.

T_i be a transaction that transfers $50 from Account A to acount B. This is defined as

T_i : read (A);
A : = A – 50;
write (A);
read (B);
B : = B + 50;
Write (B).

Applying Acid Properties

1. **Consistency :** Here consistency req. is that the sum of A and B be unchanged by the execution of transaction.
2. **Atomicity :** Let A = $1000
   B = $ 2000

If failure occures after write (A) operation was executed but buffer write (B) operation was executed. In this case, the value of account A and B reflected in the database are $1950 and $ 2000. Here we destroy $ 50 as the result of failure and sum A + B not Preserred.

To ensure atomicity, the database system keeps back (on disk) of the old value of any data on which a transaction performs a write, and, if the transaction does not complete its execution, the old values are restored to make it appear as though the transactions never-executed. Atomatic ensures by **transaction management component**.

3. Durability : The durability property guarantes not, once a transaction completes successfully, all updates that carried out on the database persist, even if there is a system failure after the transaction completes execution. To ensure durability :

1. The updates carried out by the transaction have been written to disk before the transaction completes.
2. Information about the updates carried out by the transaction and written to disk is sufficient to enable the database to reconstruct the updates when the DB system is restarted after failure. Durability Ensures by **Recovery Management Component**.

4. **ISolation :** If consistency and atomicity properties are ensured for each transaction, if several transactions are executed concurrently them operations may interleave some undesirable way resulting in an inconsistent state.

Example : Two transactions executing concenrrently may interleave some inconsistent value and A + B property is not obeyed. To solve it concurrently executing transactions is execute serially.

8.3. IMPLEMENTATION OF ATOMICITY AND DURABILITY

The recovery management component of a database system can support atomicity and durability by a variety of schemes. Consider a scheme called shadow copy scheme. This scheme, which is based on making copies of the database, called shadow copies, assumes that only one transaction is active at a time. The scheme also assumes that the database is simply a file on disk. A ppinter called db-pointer is maintained on disk, it points to the current copy of the database.

In the shadow copy scheme, a transaction that wants to update the database first creates a complete copy of the database. All updates are done on the new database copy, leaving the original copy, the shadow copy untouched. If at any point the transaction has to be aborted, the system merely deletes the new copy. The old copy of the databse has not been affected. If the transaction completes it is committed as follows. First, the operating system is asked to make sure that all pages of the new copy of the database have been written out to disk. After the O.S. has written all the pages to disk, the database system updates the pointed db.pointer to point to teh enw copy of the database the new copy then becomes the current copy of the database. The old copy is then deleted. The transaction is said to have been committed at the point where the updated dp.pointer is written to disk.

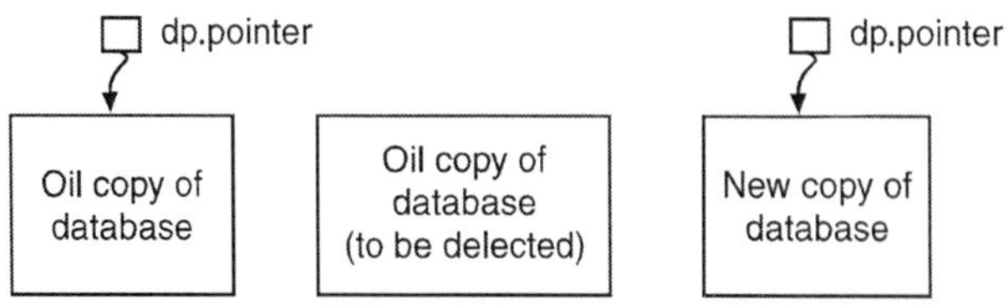

(a) Before update *(b) After update*

Shadow-copy technique for atomicity and durability

The implementation actually depends on the write to db.pointer being atomic, i.e., either all its bytes are written or none of its bytes are written. If some of the bytes of the pointer were updated by the write, but others are not, the pointer is meaningless, and neither old nor new versions of the database may be found when the system restarts. Luckily, disk systems provide atomic updates to entire blocks, or at least to a disk sector. In other words, the disk system guarantees that it will update db.pointer automatically, as long as we make sure that db.pointer lies entirely in a single sector, which we can ensure by stroing db.pointer at thebegining of a block.

Thus, the automicity and durability. Properties of transactions are ensured by the shadow-copy implementation of the recovery-management component.

8.4. IMPLEMENTATION OF ISOLATION

Since we are familiar with properties a schedule must have if it is to leave the database in a consistent state and allow transaction failures to be handled in a safe manner. Specifically, schedules that are conflict or view serializable and cascadeless satisfy these requirements.

There are various concurrency-control schemes that we can use to ensure that, even when multiple transactions are executed concurrently, only acceptable schedules are generated, regardless how the operating system time-shares resources among the transactions,

As a trivial example of a concurrency-control scheme. A transaction acquies a look on the entire database before it starts and releases the lock after it has committed. While a transaction holds a lock no other transaction is allowed to acquire the lock, and all must therefore wait for the lock to be released. As a result of the locking policy, only one transaction can execute at a time. Therefore, only serial schedules are generated. There are trivially serializable and it is easy to verify that they are cascadeless as well.

A concurrency control scheme such as this one leads to poor performance, since it forces transactions to wait for preceding transactions to finish before they can start. In other words, it provides a poor degree of concurrency.

The goal of concurrency-control schemes is to provide a high degree of concurrency, while ensuring that all schedules that can be generated are conflict or view serializable and are cascadeless.

8.5. TRANSACTION STATE

Active: It is the initial state; the transaction stays in this state while it is executing.

Partially committed : After the final statement has been executed.

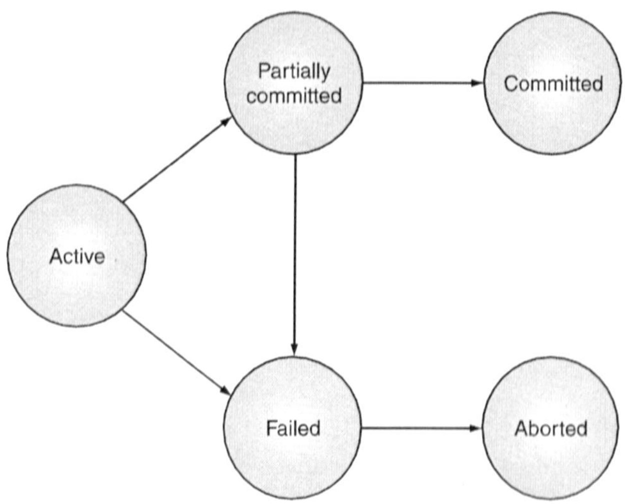

Fig. 8.1. State diagram of a transaction.

Aborted : After the transaction has been rolled back and the database has been restored to its state prior to the start of the transaction.

Committed: A transaction that completes its execution successfully is said to be committed.

Failed : After the discovery that normal execution can no longer proceed.

8.6. TRANSACTION PROCESSING SYSTEM

Transaction processing systems are the systems, which includes large size database and allows many users to use same database concurrently. For example, Railway reservation system, Banking, Credit card processing system or other similar system is called transaction processing system.

8.7. SERIALIZABILITY OF SCHEDULES

"When several transactions are executing concurrently then the order of execution of various instructions is known as a **schedule.**"

The concept of serializability of schedules is used to identify which schedules are connect when transaction executions have interleaving of their operations in the schedules.

Types of schedules

1. Serial schedules.
2. Non-serial schedules.
3. Conflict schedules.
4. View schedules.

Serial Schedules

T_1	T_2
read-item (X) ;	
X : X-N ;	
Write-item (X);	
read-item (Y) ;	
Y : Y + N ;	
write-item (Y) ;	
	read-item (X) ;
	X : X + M ;
	write-item (X) ;

Schedule A

T_1	T_2
	read-item (X) ;
	X : X + M ;
	write-item (X) ;
read-item (X) ;	
X : X + N ;	
write-item (X) ;	
read-item (Y) ;	
Y : = Y – N ;	
write-item (Y) ;	

Schedule B

Two schedules *A* and *B* are called serial schedule if the operations of each transaction are executed consecutively, without any interleaved operations from the other transaction. In a serial schedule, entire transaction are performed in serial order T_1 and then T_2 or T_2 and then T_1.

Non-serial Schedules

Two schedules C and D are called non-serial schedule if the operations of each transaction are executed non-consecutively with interleaved operations from the other transaction.

T_1	T_2	T_1	T_2
read-item (X) ;			
X : X-N ;			
	read-item (X) ;		
	X : X + M ;		
Write-item (X);			
read-item (Y) ;			
	write-item (X) ;		
Y : Y + N ;			
write-item (Y) ;			

Conflict Schedules : The graph which contains no cycle is called **conflict serializable schedules**.

If the precedence graph for *S* has a cycle, then schedule *S* is not conflict serializable. "If graph contains no cycle, then the schedule *S* is conflict serializable."

8.8. TESTING OF SERIALIZABILITY

For testing of serializability the simple and efficient method is to construct a directed graph, called a precedence graph from *S*.

$$G = (V, E)$$

The set of vertices consists of all the transactions participating in the schedule. The set of edges consists of all edges $T_i \rightarrow T_j$, for which one of three conditions holds :

1. T_i executes write (Q) before T_j executes read (Q).
2. T_i executes read (Q) before T_j executes write (Q).
3. T_i executes write (Q) before T_j executes write (Q).

For example, *S* :

T_1	T_2
Read (A) A : A-50	
	read (A) temp : = A × 0.1 A : = A-temp write (A) read (B)
write (A) Read (B) B : = B + 50 write (B)	
	B : = B + temp write (B)

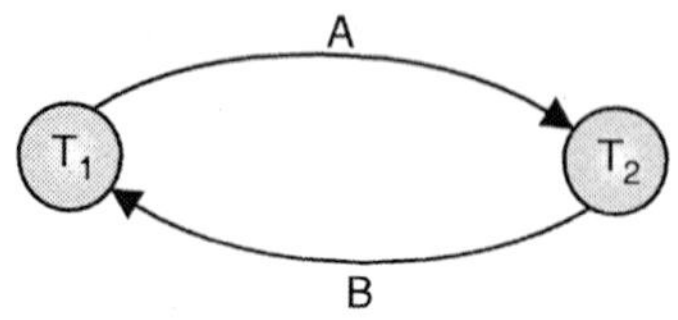

Fig. 8.2. Precedence graph for above schedule.

For example 2 :

T_1	T_2	T_3
Read (X)		
	Write (X)	
Write (X)		
		Write (X)

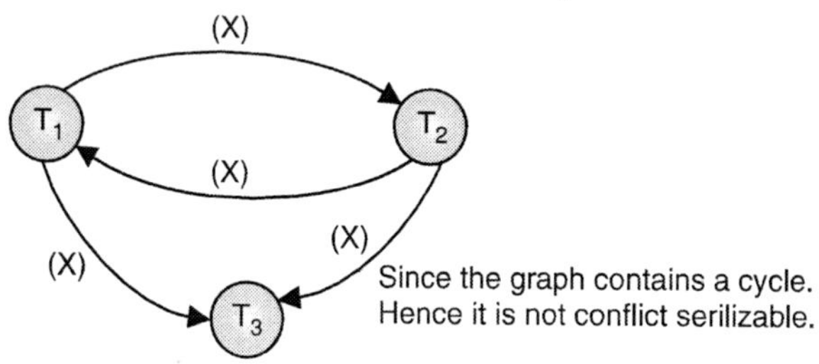

Fig. 8.3.

Since the graph contains a cycle. Hence it is not conflict serializable.

For example 3 :

T_1	T_2	T_3
$R_1(X)$		
		$R_3(X)$
$W_1(X)$		
	$R_2(X)$	
		$W_3(X)$

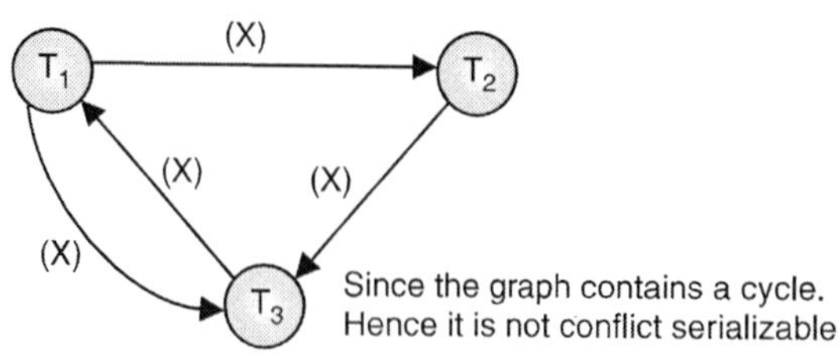

Fig. 8.4.

Since the graph contains the cycle. Hence it is not conflict serializable.

For example 4 :

T_1	T_2	T_3
$R_1(X)$		
		$R_3(X)$
		$W_3(X)$
$W_1(X)$		
	$R_2(X)$	

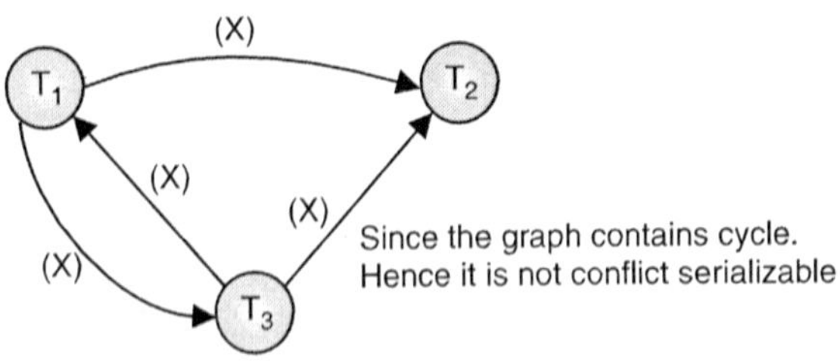

Fig. 8.5.

Since the graph contains cycle. Hence it is not conflict serializable.

For example 5 :

T_1	T_2	T_3
		$R_3(X)$
	$R_2(X)$	
		$W_3(X)$
$R_1(X)$		
$W_1(X)$		

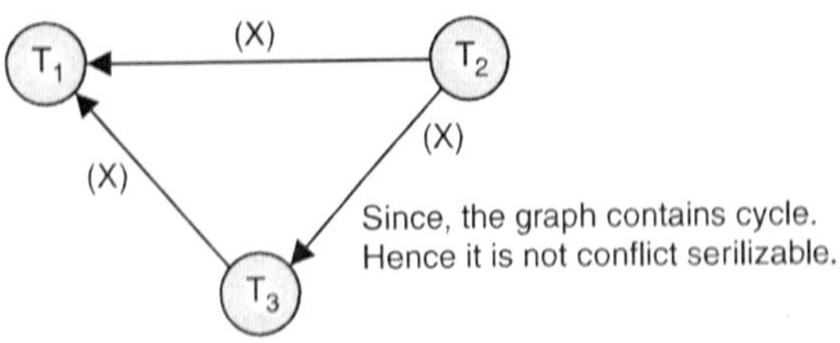

Fig. 8.6.

Since the graph contains cycle. Hence it is not conflict serializable.

For example S_1 : **Draw P.G.**

T_1	T_2	T_3
$R_1(X)$		
	$R_2(Z)$	
$R_1(Z)$		
		$R_3(X)$
		$R_3(Y)$
		$W_3(X)$
	$R_2(Y)$	
	$W_2(Z)$	
	$W_2(Y)$	

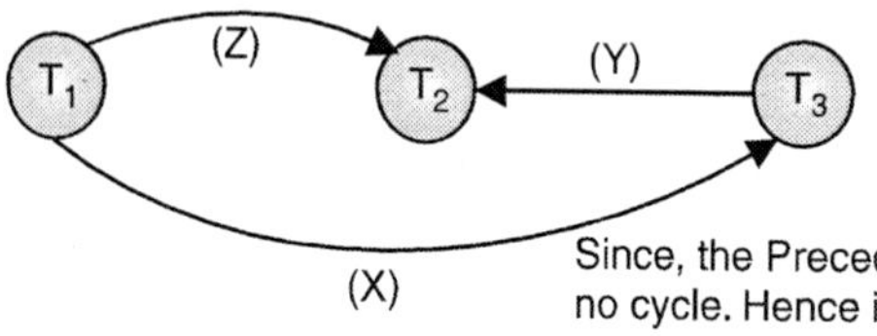

Since, the Precedence graph contains no cycle. Hence it is conflict serializable.

Fig. 8.7.

For example S_2 :

T_1	T_2	T_3
$R_1(X)$		
	$R_2(Z)$	
		$R_3(X)$
$R_1(Z)$		
	$R_2(Y)$	
		$R_3(Y)$
$W_1(X)$		
	$W_2(Z)$	
		$W_3(Y)$
	$W_2(Y)$	

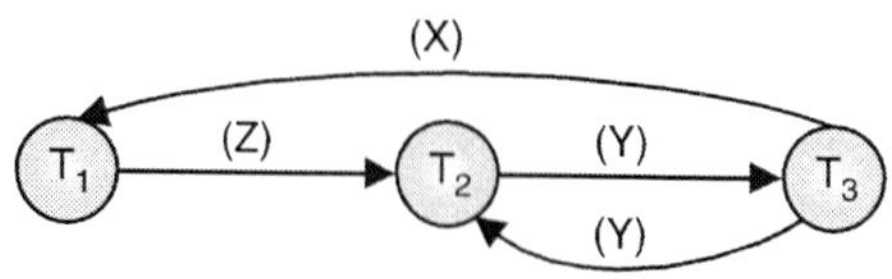

Since the P.G contains cycle, Hence it is not conflict serializable

Fig. 8.8.

8.8.1. Conflict Serializability

Consider a schedule S in which there are two consecutive instructions I_i and I_j of the transactions T_i and T_j respectively $(i \neq j)$

Now we identified three rules that show that if two operations in a schedule satisfy all these three conditions then operations are said to conflict :

1. They belong to different transactions.
2. They access the same item.
3. At least one of the operations is a write-item.

Thus, only in the case where both I_i and I_j are read instructions does the relative order if their execution not matter.

We say that I_i and I_j conflict if they are operations by different transactions on the same data item, and at least one of these instructions is a write operation.

If a schedule S can be transferred into a schedule S' by a series of swaps of non-conflicting instructions, we say that S and S' are conflict equivalent.

The concept of conflict equivalence leads to the concept of conflict serializability. We say that a schedule S is conflict serializability if it is conflict equivalent to a serial schedule.

8.8.2. View Serializability

Consider two schedules S_1 and S_2 where same set of instructions participate in both schedules. The schedules S_1 and S_2 are said to be view equivalent if the following conditions are met :

1. For each data item X, if transaction T_i reads the initial value of X in schedule S_1, the transaction T_i must, in schedule S_2 also read the initial value of X.
2. For each data item X, if transaction T_i executes read (A) in schedule S_1 and the value was produced by transaction T_j (if any), then transaction T_i must in schedule S_2 also read the value of X that was produced by transaction T_j.
3. For each data item X, the transaction (if any) that performs the final write (X) operation in schedule S_1 must perform the final write (X) operation schedule S_2.

The concept of view equivalence leads to the concept of view serializability. We say that a schedule S is view serializability if it is view equivalence to a serial schedule.

Every conflict-serializability schedule is also view serializable, but there are view serializable schedule that are not conflict serializable.

8.9. RECOVERY FROM TRANSACTION FAILURES

A computer, like any other device, is subject to failure. There may be several cause of such failure, including power failure, disk crash, software errors etc. In each of these cases, information pertaining to database system is lost.

A recovery process is an integral part of a database system, which is responsible for detection of failures and recovery of database.

There are two types of failures :

1. **Loss of Volatile Storage :** It can be recovered using various methods. example, log based recovery, Buffer management, checkpoints and shadow paging techniques.

2. **Non-Volatile Storage Loss :** The most popular and economical scheme is to dump the entire database to a stable storage periodically. For example, we may dump the database to one or more magnetic tapes. If a failure occurs the most recent dump is used in restoring the database.

There are various techniques of recovery.

8.9.1. Cascading Rollback

The phenomenon, in which a single transaction failure leads to a series of transaction rollback, is known as "**Cascading Rollback.**"

For example :

T_1	T_2	T_3
Read (X)		
Read (Y)		
$Z = X + Y$		
Write (Z)		
	Read (Z)	
	Write (Z)	
		Read (Z)

Transaction T_1, writes a value of Z, that is read by transaction T_2. Transaction T_2 writes a value of Z is read by transaction T_3.

Suppose that, at this point, T_1 fails, T_1 must be rolledback. Since T_2 is dependent on T_1, T_2 must be rolled back. Since T_3 is dependent on T_2, T_3 must also be rolled back.

8.9.2. Recoverable Schedules

T_1	T_2
Read (X)	
Write (X)	
	Read (X)
Read (Y)	

In this schedule T_2 transaction performs only one transaction read (X).

Suppose that the system allows T_2 to commit immediately after excluding the read (X) instruction. It means T_2 commits before T_1.

Now suppose that T_1 fails before it commits. Since T_2 has already read the data item (X) written by T_1. So it is necessary to abort T_2 to ensure transaction atomicity.

However, T_2 has already committed and it cannot be aborted. Now we have arrived in a situation where it is not possible to recover correctly from the failure of T_1.

Hence transaction processing system must ensure that it is possible to recover from the failure of any active transaction.

8.9.3. Log Based Recovery

To keep track of the database transactions, the DBMS maintains special files called log files or journals that contain information about all updates. The log file contains information like transaction identifier, type of the log record etc.

Since the information contained in the log files are critical for the database recovery, two separate copies are maintained. In the past the log files were stored on magnetic tapes. But nowadays, DBMS are expected to recover from minor failures very quickly and so the log files are stored online on a fast direct access storage devices (DASD).

Since the online log files are also prone to failures, the log files are periodically archived and are stored in off-line storage. These log files are called "**archive log files.**"

Log

The most widely used structure for recording database modifications is the log. The log is a sequence of log records, recording all the update activities in the database. There are several types of log records. An update log describes a single database write. It has these fields:

#Transaction identifier is the unique identifier of the transaction that performed the write operation.

#Data item identifier is the unique identifier of the data item written. Typically, it is the location on disk of the data item.

#Old value is the value of the data item prior to the write.

#New value is the value that the data item will have after the writer.

Other special log records exist to record significant events during transaction processing, such as that start of a transaction and the commit or abort of a transaction, we denote the various type of log records as:

- $<T_1$ start.$>$ Transaction T_i has started.
- $< T_i X_j, V_1, V_2 >$. Transaction T_i has performed a write on data item X_j. X_j had value V_1 before the write and will have V_2 after the write.
- $< T_i$ commit$>$ Transaction T_i has committed.
- $< T_i$ abort$>$ Transaction T_i has aborted.

It is of two types : (*i*) Deferred D.M.
(*ii*) Immediate D.M.

(*i*) Deferred Database Modification : The deferred modification technique ensures transaction atomicity by recording all database modifications in the log, but deferring the execution of all write operations of a transaction until the transaction partially commits. Recall that a transaction is said to be partially committed once the final action of the transaction has been executed. The version of the deferred modification technique that we describe in this section assumes that transactions are executed serially.

Suppose that these transactions are executed serially, in the order T_0 followed by T_1, and that the values of accounts A, B and C before the execution took place were \$1000, \$2000, and \$700, respectively. The portion of the log containing the relevant information on these two transactions appears in figure. There are various orders in which the actual outputs can take place to both the database system and the log as a result of the execution of T_0 and T_1.

<T_0 start>
< T_0 A,950>
<T_0 B, 2050>
<T_0 commit>
<T_1 start>
<T_1 C, 600>
<T_1 commit>

***(ii)* Immediate Database Modification:** The immediate modification technique allows database modifications to be output to the database while the transaction is still in the active state. Data modifications written by active transactions are called uncommitted modifications. In the event of a crash or a transaction failure, the system must use the old value field of the log records described to restore the modified data items to the value they had prior to the start of the transaction. The undo operation described next, accomplishes this restoration.

Log	Database
< T_0 start >	
< T_0 A, 1000, 950 >	
< T_0 B, 2000, 2050 >	A = 950
< T_0 commit >	B=2050
< T1 start>	
< T_1 C, 700, 600 >	C=600
< T_0 commit >	

This figure shows one possible order in which the actual outputs took place in both the database system and the log as a result of the execution of T_0 and T_1.

8.9.4. Checkpoints

A checkpoint is a point of synchronisation between the database and the transaction log file. All buffers are force-written to secondary storage at the checkpoint. Checkpoints are also called syncpoints or save points.

Checkpoints are schedules at pre-determined interval and involves operations like writing all log records in main memory to secondary storage.

If the transactions are executed serially, when a failure occurs, we check the log file to find the transaction that started before the last checkpoint.

When a systems failure occurs, we must consult the log to determine those transactions that need to be redone and those that need to be undone. In principle, we need to search the entire log to determine this information. There are two major difficulties with this approach.

- The search process is time-consuming.
- Most of the transactions that need to be redone have already written their updates into the DB. Although redoing them will cause no harm, it will nevertheless cause recovery to take longer.

To reduce these types of overhead, *Check points* are introduced. During execution, the system maintains the log. In addition, the system periodically performs checkpoints, which require the following sequence of actions to take place.

- Output onto stable storage all log records currently residing in main memory.
- Output to the disk all modified buffer blocks.
- Output onto stable storage a log record < check point >.

Transactions are not allowed to perform any update actions, such as writing to a buffer block or writing a log record, while a checkpoint is in progress.

The presence of a <checkpoint> record in the log allows the systems to streamline its recovery procedure.

Consider a transaction T_i that committed prior to the checkpoint. For such a transaction; the <T_i commit> record appears in the log before the <checkpoint> record. Any DB modifications made by T_i most have been written to the DB either prior to the checkpoint tot he checkpoint or as part of the checkpoint itself. Thus, at recovery time, there is no need to perform a *redo* operation of T_i.

This observation allows us to refine our previous recovery schemes (Assume transactions are run serially). After a failures has occurred, the recovery scheme examines the log to determine the most recent transaction T_i that started executing before the most recent checkpoint took place. It can find such a transaction by searching the log backward, from the end of the log, until it finds the first <checkpoint> record; then it continues the search backward until it finds the next <T_i start> record. This record identifies a transaction T_i.

One T_i has been identified, the redo and undo operations need to be applied to only T_i and all transaction T_j that started executing after T_i. Let denote these transactions by the set T. The remainder (earlier part) of the log can be ignored, and can be erased whenever desired. The exact recovery operations to be performed depend on whether the *immediate-modification technique* or the *defined modification technique* is being used.

If the Immediate Modification Technique is employed, the recovery operations that are required are as :

- For all transactions T_K in T that have no <T_K commit> record in the log, execute undo (T_K)
- For all transactions T_K in T such that record <T_K commit> appears in the log, execute redo (T_K).

Obviously, the undo operation does not need to be applied when the defined modification technique is being employed.

Illustration : Consider set of transactions $\{T_0, T_1, ..., T_{100}\}$ executed in sequence. Suppose the most recent checkpoint took place during the execution of transaction T_{67}. Thus, only transactions T_{67}, T_{68}, ..., T_{100} need to be considered during the recovery scheme.

Each of them needs to be redone if it has committed. Otherwise, it need to be *undone*.

8.9.5. Backup Mechanism

The backup copy of the database can be used to recover the database in the event that the database has been damaged or destroyed. A backup can be a complete copy of the entire database or an incremental copy. An incremental backup consist only of modifications made since the last complete or incremental backup.

8.9.6. Shadow Paging

The shadow paging scheme does not require the use of a log file in a single user environment, though a log is needed in multi-user environment for concurrency control. Shadow paging considers the database to be made up of a number of fixed-size disk pages or disk blocks for recovery purposes.

The shadow paging technique maintain two directories (two for each database page) during the life of a transaction—a current directory and a shadow directory. When the transaction starts, the two directories are the same. The shadow directory is saved to the disk and the current directory is used by the transaction. The shadow directory is never changed and is used to restore the database in case of a failure.

During the transaction execution, the shadow directory is never modified.

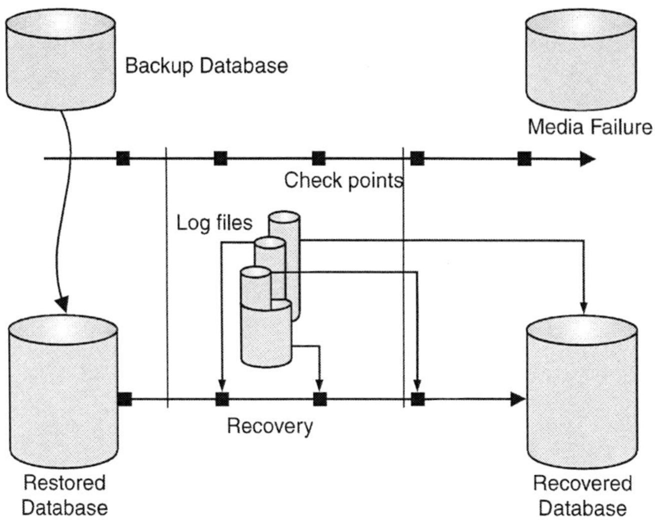

Fig. 8.9. Restoring and recovering a database

SOLVED EXAMPLES

Example 8.1. Which of the following schedule is a conflict serializable ? For each conflict serializable schedule, determine its equivalent serial schedule.

Solution : (*a*) $r_1(X)$; $r_3(X)$; $W_1(X)$; $r_2(X)$; $W_3(X)$

T_1	T_2	T_3
$r(X)$		
		$r(X)$
$W(X)$		
	$r(X)$	
		$W(X)$

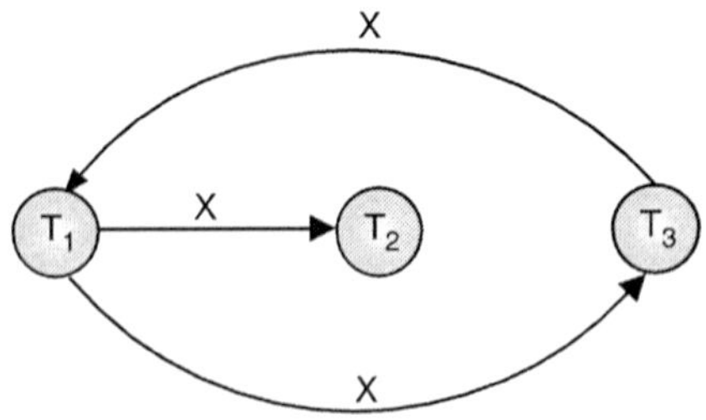

Fig. 8.10.

Since the precedence graph contains cycle. Hence it is not conflict serializable. Hence its equivalent schedule can not be determine.

(*b*) $r_1(X)$; $r_3(X)$; $W_3(X)$; $W_1(X)$; $r_2(X)$

T_1	T_2	T_3
$r(X)$		
		$r(X)$
		$W(X)$
$W(X)$		
	$r(X)$	

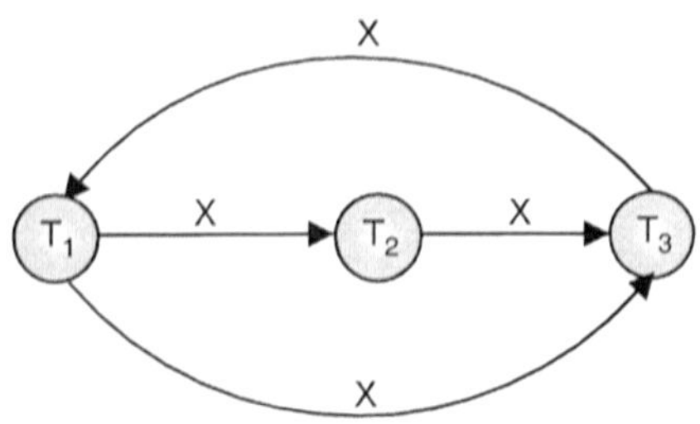

Fig. 8.11.

Since the precedence graph contains cycle. Hence it is not conflict serializable. Therefore its serial schedule cannot be determined.

(c) $r_3(X)$; $r_2(X)$; $W_3(X)$; $r_1(X)$; $W_1(X)$

T_1	T_2	T_3
		$r(X)$
	$r(X)$	
		$W(X)$
$r(X)$		
$W(X)$		

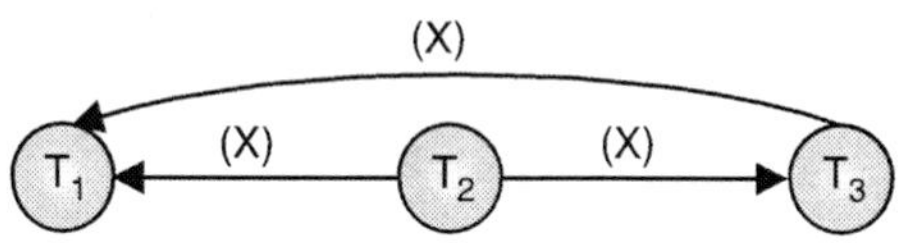

Fig. 8.12.

Since the precedence graph contains no cycle. Hence it is not conflict serializable and its equivalent serial schedule is as follows :

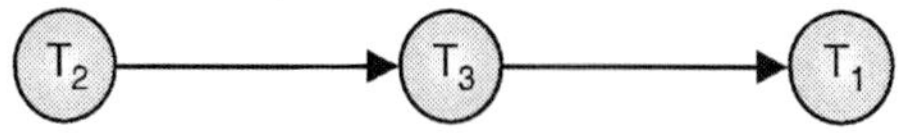

Fig. 8.13.

T_2	T_3	T_1
$r(X)$		
	$r(X)$	
	$W(X)$	
		$r(X)$
		$W(X)$

Example 8.2. Check whether the given schedule is conflict serializable or not :

(a) SB : $r_1(Q)$, $W_2(Q)$, $r_1(Q)$

(a) SC : $r_1(Q)$, $W_2(Q)$, $W_1(Q)$, $W_3(Q)$

Solution : (a)

T_1	T_2	
$r(Q)$		
	$W(Q)$	
$r(Q)$		

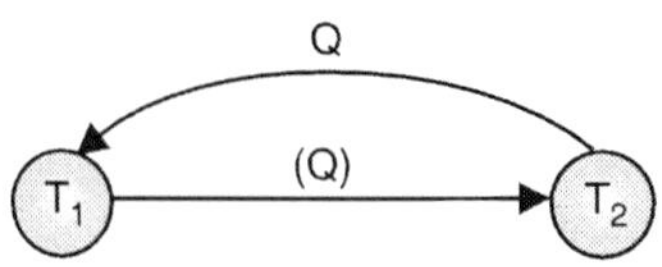

Fig. 8.14.

Since the precedence graph contains cycle. Hence it is not conflict serializable sechedule.

Solution : (*b*) SC : $r_1(Q)$, $W_2(Q)$, $W_1(Q)$, $W_3(Q)$

T_1	T_2	T_3
$r(Q)$		
	$W(Q)$	
$W(Q)$		
		$W(Q)$

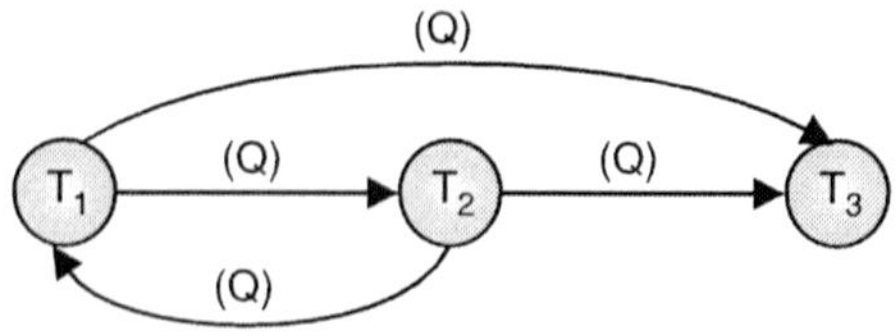

Fig. 8.15.

Since the precedence graph contains cycle. Hence it is not conflict serializable sechedule.

Summary

- Collections of operations that form a single logical unit of work are called transaction.
- Transactions have four basic properties, which are called ACID properties.
- Transaction processing system are the systems, which includes large size database and allows many users to use same database concurrently.
- The phenomenon in which a single transaction failure leads to a series of transaction rollback is known as "Cascading Rollback".
- Checkpoints is a point of synchronization between the database and the transaction log file.
- When several transactions are executing concurrently then the order of execution of various instruction is known as a "Schedule".
- The precedence graph which contains no cycle is called "Conflict serializable".
- A system is said to be in deadlock state when in a set of transaction, every transaction is waiting for another transaction to finish its operations.
- Deadlock can be described precisely in terms of a directed graph called a wait for graph. This graph consist of a pair $G = (V_1\ E)$, where V is a set of vertices and E is a set of edges.
- When a detection algorithm determines that a deadlock exists the system must recover from the deadlock. The most common solution is to rollback one or more transactions to break the deadlock.

Selected Exercises

1. What do you mean by transaction? Explain the transaction states.
2. What are the ACID properties? Explain in brief.
3. Explain the transaction processing system in DDBMS with figure.

4. What are various techniques of Recovery from Transaction failures ?
5. Explain log based Recovery and Checkpoints.
6. What do you meant by deadlock ?
7. Which of the following schedule is conflict serializable ? For each serializable schedule determine the equivalent serial schedule.
   (i) $r_1(X)$; $r_3(X)$; $w_1(X)$; $r_2(X)$; $w_3(X)$
   (ii) $r_1(X)$; $r_3(X)$; $w_3(X)$; $w_1(X)$; $r_2(X)$
   (iii) $r_3(X)$; $r_2(X)$; $w_3(X)$; $r_1(X)$; $w_1(X)$
8. Consider the three transactions T_1, T_2 and T_3 and the schedules S_1 and S_2 given below. Draw the serializability (procedure) graph for S_1 and S_2.

   $T_1 : r_1(X) ; r_1(Z) ; w_1(X) ;$

   $T_2 : r_2(Z) ; r_2(Y) ; w_2(Z) ; w_2(Y) ;$

   $T_3 : r_3(X) ; r_3(Y) ; w_3(Y) ;$

   $S_1 : r_1(X) ; r_2(Z) ; r_1(Z) ; r_3(X) ; r_3(Y) ; w_3(Y) ; r_2(Y) ; w_2(Z) ; w_2(Y) ;$
   $S_2 : r_1(X) ; r_2(Z) ; r_1(X) ; r_1(Z) ; r_2(Y) ; r_3(Y) ; w_1(X) ; w_2(Z) ; w_3(Y) ; w_2(Y).$
9. $r_1(X)$, $w_1(X)$, $r_2(X)$, $w_2(X)$,
   $r_1(B)$, $w_1(B)$, $r_2(B)$, $w_2(B)$.
   $S_B : r_1(\theta), w_2(\theta), r_1(\theta)$
   $S_C : r_1(\theta), w_2(\theta), w_1(\theta), w_3(\theta)$
   Check whether the given schedule is conflict serializable or not.
10. Explain serializability of schedule and also its types.
11. What do you meant by deadlock handling? Explain in brief.
12. Explain the conflict and view serializable schedule in details.

❒❒❒

Chapter

9 Concurrency Control Techniques

Inside this Chapter

9.1. CONCURRENCY CONTROL

Concurrency control is the management of concurrent transaction execution. DBMS implement concurrency control techniques to ensure serializability and isolation of transaction in order to guard the consistency and integrity of the database.

Concurrency control can be performed by the DBMS, with various methods such as locking methods, timestamp methods, validation based techniques and optimistic methods.

Concurrency control is the process of managing simultaneous operations (queries, updates, inserts, deletes) on the database without having them interfere with one another.

The objective of concurrency control is similar to the objecitve of multi-user computer system where many users can perform, different operations at the same time. In these systesm the different users are allowed to perform different actions simultaneously due to concept called multi-programming.

9.2. LOCKING TECHNIQUES FOR CONCURRENCY CONTROL

Locking is a procedure used to control concurrent access to data. When one transaction is accessing the database, a lock may be deny access to other transactions to prevent incorrect results.

Locking is one of the most widely used mechanism to ensure serializability. The working principle of locking is as follows : A transaction must obtain a read or write lock on a data item before it can perform a read or write operation.

The basic rules for locking are :

1. If a transaction has a read lock on a data item, it can read the item but not update it.
2. If a transaction has a read lock on a data item, other transactions can obtain a read lock on the data item, but not write locks.
3. If a transaction has a write lock on a data item. It can both read and update the data item.
4. If a transaction has a write lock on a data item, then other transactions cannot obtain either a read lock or a write lock on the data item.

A most commonly used technique to control the concurrency is based on the concept of locking data items. A lock on a data item specifies the status of the item with respect to possible operations like read and write that can be applied to it. If a transaction *T* wants to modify a data item, a lock must be associated with that item. If any other transaction, Say T_2 already accesses the same data item, T_1 has to wait until T_2 gives up the lock. Otherwise T_1 locks the data item.

9.3. MODE OF LOCKING

There are mainly two modes in which a data item may be locked :

1. Shared lock
2. Exclusive lock.

9.3.1. Shared Lock

If a transaction T_i has obtained a shared-mode lock on item *Q*, then T_i can read but cannot write *Q*. It is denoted by (*S*).

9.3.2. Exclusive Lock

If a transaction T_j has obtained an exclusive-mode lock on item *O*, then T_j can both read and write *O*. It is denoted by (*X*).

Now depending upon the operation performed on data item X every transaction should request an appropriate lock on X. A transaction requests a shared lock on data item X by executing the – S(X) instruction. Similarly an exclusive lock is requested by the X(X) instruction. A data item X may be unlocked by Unlock (X) instruction.

Based on these locking nodes, we can define a compatibility function as :

Let A & B be two arbitrary locking modes. Suppose a transaction T_1 holds a A lock on a data item X and another transaction T_2 requests a B lock on the same item. Then if transaction T_2 gets the permission to lock X in B mode then we say that B mode is compatible or non-conflicting to A mode. After finishing the operations on a data item, the transaction unlocks the data item.

In order to access a data item :

(i) Transaction T_1 must first lock it.

If data item is already locked by other transaction T_2 which is not compatible, then T_1 most wait until locks have been released and unlocked.

(ii) Transactions must hold a lock on data item as long as it accesses it.

(iii) And then unlock it.

A matrix gives below can show the compatibility between two lock modes :

	S	X
S	T	F
X	F	F

Here T means that the modes are compatible to each other. Shared mode locking is compatible with another shared mode of locking but not with exclusive mode of locking. Whereas exclusive mode of locking is neither compatible with exclusive mode nor shared mode of locking.

For example : Consider X = 100, Y = 50, N = 4 and M = 5. At the end of transaction X + Y should be equal to 150. The figure is :

T_1	
X(X)	
Read (X)	X = 100
X = X – N	X = 100 – 4
Write (X)	X = 96
Unlock (X)	
X(Y)	
Read (Y)	Y = 50
Y = Y + N	Y = 50 + 4
Write (Y)	Y = 54
Unlock (Y)	
X + Y = 96 + 54 = 150	

T_2	
X(X)	
Read (X)	X = 96
X = X + M	X = 96 + 5
Write (X)	X = 101
Unlock (X)	
X(Y)	
Read (Y)	Y = 54
Y = Y – M	Y = 54 – 5
Write (Y)	Y = 49
Unlock (Y)	
X + Y = 96 + 54 = 150	

If these two transaction T_1 and T_2 are executed serially, either in the order T_1, T_2 or the order T_2, T_1 In both the cases we will get the same result X + Y = 150, which is correct. If, however these transactions are executed concurrently, then there are many possibilities how these transactions will be executed. A schedule is given for two transactions in the figure below. In this case the transaction T_2 would give the wrong result, which will also affect the result of transaction T_1. The reason for this is that the transaction T_1 has unlocked data item X too early.

T_1		T_2	
X(X)			
Read (X)	X = 100		
X = X – N	X = 100 – 4		
Write (X)	X = 96		
Unlock (X)			
		X(X)	
		Read (X)	X = 96
		Y = X + M	X = 96 + 5
		Write (X)	X = 101
		Unlock (X)	
X(Y)		X(Y)	
Read (Y)	Y = 45	Read (Y)	Y = 50
Y = Y + N	Y = 45 + 4	Y = Y – M	Y = 50 – 5
Write (Y)	Y = 49	Write (Y)	Y = 45
Unlock (Y)		Unlock (Y)	
X + Y = 96 + 44 = 145		X + Y = 101 + 45 = 146	

Suppose now that unlocking is delayed to the end of transaction as :

T_1		T_2	
X(X)			
Read (X)	X = 100		
X = X – N	X = 100 – 4		
Write (X)	X = 96		
		X(X)	
		Read (X)	X = 96
		Y = X + M	X = 96 + 5
		Write (X)	X = 101
		Unlock (X)	
X(Y)		X(Y)	
Read (Y)	Y = 50	Read (Y)	Y = 54
Y = Y + N	Y = 50 + 4	Y = Y – M	Y = 54 – 5
Write (Y)	Y = 54	Write (Y)	Y = 49
Unlock (Y)		Unlock (Y)	
Unlock (X)			
X + Y = 96 + 54 = 150		X + Y = 101 + 49 = 150	

In this case we have delayed Unlock (X) instruction to the end of transaction. Due to this delay T_2 will not be able to access data item X till the end of execution of T_1 and we would get the correct results.

9.4. THE TWO-PHASE LOCKING PROTOCOL

(UPTU, 2004, 2006, 2008)(Imp.)

The two-phase locking protocol ensures serializability. This protocol requires that each transaction issue lock and unlock requests in two phase :

1. Growing phase
2. Shrinking phase

1. Growing Phase : In this phase a transaction may obtain locks but may not release any lock. Here the no. of locks increases from zero to the maximum for the transaction.

2. Shrinking Phase : In this phase a transaction may release locks but may not obtain any new locks. Here the no. of locks decreases from maximum to zero.

There are three types of two-phase locking protocol :

(*a*) Conservative Two-phase Locking.

(*b*) Strict Two-phase Locking.

(*c*) Rigorous Two-Phase Locking.

9.4.1. Conservative Two-Phase Locking

It requires a transaction to pre-declare all the locks it requires before its execution.

If any of pre-declared items cannot be locked the transaction waits until all item are locked.

The locking protocol is deadlock free but it is difficult to use in practice.

9.4.2. Strict Two-Phase Locking

In this protocol a transaction does not release any of its write lock until it commits or aborts.

9.4.3. Rigorous Two-Phase Locking

In this protocol, a transaction does not release any of write lock and read lock until it commits or aborts.

9.5. TIME STAMPING PROTOCOLS FOR CONCURRENCY CONTROL

(UPTU, 2006-07, 2010) (05 Marks)

Time stamping is a concurrency control protocol in which the fundamental goal is to order transactions globally in such a way that older transactions get priority in the event of a conflict. In the timestamping method, if a transaction attempts to read or write a data item, then a read or write operation is allowed only if the last updata on that data item was carried out by an older transacton. Otherwise the transaction, requesting the read or write is re-started and given a new timestamp. The restarted transaction is assigned a new timestamp to prevent it from continuously aborting and restarting.

In this method sequence of transaction is selected in advance by using the timestamping concept. We associate a unique, fixed non-decreasing numbers called the timestamp to each transaction in the system. This number can be the system clock value, *i.e.*, when the transaction enters the system the current value of the clock is assigned to that transaction as the timestamp value. We can also use a logical counter that is incremented after a new transaction enters the system. A transaction with a smaller timestamp value is considered to be an adder transaction than another transaction with a larger timestamp value. In other words, if a transaction T_1 has been assigned a timestamp S_1 and a new transaction T_2 enters the system with a timestamp value S_2 then $S_1 < S_2$ and system will allow the transaction T_1 to run first followed by the transaction T_2.

If any confliction is found between transaction using timestamp based system, one of the conflicting transactions is rolled back.

- When a transaction having smaller timestamp value tries to access a data value that is written by the transaction having larger timestamp value.
- When a transaction with smaller timestamp value tries to update a data value already read or written by a transaction having larger timestamp value.

W_TS(X) : Let W_TS be the timestamp value of most recent write of any transaction (means having the largest timestamp value). That is W_TS (X) = TS (T) when T is the youngest transaction that has written X successfully.

R_TS(X) : R_TS be the largest timestamp value of the transaction having successfully executed read operation. That is R_TS(X) = TS(T) where T is the youngest transaction that has read X successfully. Every time when a new r/w instruction is executed these values are updated. In order to ensure that any conflicting r/w operations are executed in the timestamping, following rules are followed :

(A) Suppose transaction T issues a read operation on data item X.

(*i*) If TS(T) < W_TS (X) then abort and rollback the transaction because T needs to read a value of X which has already been over written by some younger transaction with timestamp greater than TS(T).

(*ii*) If first condition does not hold, *i.e.*, W_TS(X) <= TS(T) then read operation on X is executed and R_TS(X) is set to the larger value of TS(T) and current R_TS(X).

(B) Suppose transaction T issues a write operation :

(*i*) If TS(T) < R_TS(X) then T is producing a value to X which was previously needed by a transaction and having a larger stamp value than T. Hence the write operation is rejected and transaction T is rolled back.

(*ii*) If TS(T) < W_TS(X) then T is trying to write an obsolete value into X. Hence write operation is rejected and transaction T is rolled-back.

(*iii*) Otherwise write operation is executed and set W_TS(X) to TS(T).

* Write a transaction is aborted and rolled back a new timestamp value is assigned to T and T is re-started.

To better understand timestamp based protocol, consider an example : Suppose there are two transactions T_1 and T_2 which transfer some amount from account A to account B and then display the result in another transaction T_3 as follows :

T_1	T_2	T_3
Read A	Read A	Read A
A = A – 100	A = A – 250	Read B
Write A	Write A	
Read B	Read B	Display
B = B + 100	B = B + 250	A + B
Write B	Write B	

Suppose the timestamp values of these transactions are 10, 20 and 30 respectively. A schedule that follows the time stamping protocol is show in figure below. Here last 2 columns read and write time stamp values (R_TS and W_TS) of A & B.

T_1(10)	T_2 (20)	T_3 (30)	A(R_TS = 0,	B(R_TS = 0
			W_TS = 0	W_TS = 0
Read A			R_TS = 0	
A = A – 100				
Write A			W_TS = 10	
Read B	Read A		R_TS = 20	R_TS = 10
B = B + 100	A = A – 250			
Write B	Write A		W_TS = 20	W_TS = 10
	Read B	Read A	R_TS = 30	
	B = B + 250			
	Write B			W_TS = 20
		Read B		R_TS = 30
		Display A + B		

9.6. DEADLOCK HANDLING

A system is said to be in deadlock state when in a set of transactions, every transaction is waiting for another transaction to finish its operations.

A deadlock is a state of statement that may result when two or more transactions are each waiting for locks held by the other to be released.

There are two principal methods for dealing with deadlock problems :

1. Deadlock prevention protocol
2. Deadlock detection and recovery process.

1. Deadlock Prevention Protocol : This protocol ensures that the system will not go into deadlock state. There are different methods that can be used for deadlock prevention :

(*a*) The simplest scheme requires that each transaction locks all its data item before it starts execution.

(*b*) Another method for the prevention of deadlock is to impose a partial ordering of all data items and require that a transaction can lock a data item only in the order specified by partial order.

(*c*) Another approach for the prevention of deadlock is to use pre-emption and transaction rollbacks.

2. Deadlock Detection and Recovery : If a system does not imply some protocol that ensures deadlock freedom, then a detection and recovery scheme must be used.

9.7. DEADLOCK DETECTION

Deadlock can be described precisely in terms of a directed graph called a wait for graph. This graph consist of a pair $G = (V, E)$, where V is a set of vertices and E is a set of edges. Each element in the set E of edges is an ordered pair $T_i \rightarrow T_j$. If $T_i \rightarrow T_j$ is in E, then there is a directed edge from transaction $T_i + T_j$, implying that transaction T_i is waiting for transaction T_j to release a data item that it needs.

A deadlock exists in the system if and only if the wait for graph contains a cycle. Each transaction involved in the cycle is said to be deadlocked.

This wait for graph contains a cycle. Hence T_{26}, T_{27}, T_{28} are all deadlocked.

9.8. RECOVERY FROM DEADLOCK

When a detection algorithm determines that a deadlock exists, the system must recover from the deadlock. The most common solution is to rollback one or more transactions to break the deadlock. There actions need to be taken :

1. **Selection of a Victim :** In this we determine which transaction (or transactions) to roll back to break the deadlock. We should rollback those transactions that will incur the minimum cost.
2. **Rollback :** Once we have decided that a particular transaction must be rolled back, we must determine how far this transaction should be rolled back. The simplest solution is a "total rollback". Abort the transaction and then restart it. However, it is more effective to roll back the transaction only as far as necessary to break the deadlock. Such partial rollback requires the system to maintain additional information about the state of all the running transactions.
3. **Starvation :** In a system where selection of transactions, for rollback, is based on the cost factor, it may happen that the some transaction are always picked up.

9.9. FAILURE CLASSIFICATION

There are various types of failure that may occur in a system, each of which needs to be dealt in a different manner. The simplest type of failure is one that does not result in the loss of information in the system. The failures that are more difficult to deal with are those that result in loss of information.

Transaction Failure

There are two types of errors that may cause a transaction to fail :

(*i*) Logical error: The transaction can no longer continue with its normal execution because of some internal condition, such as bad input, data not found, overflow, or resource limit exceeded.

(*ii*) System error: The system has entered an undesirable state (for e,g, deadlock), as a result of which a transaction cannot continue with its normal execution. The transaction, however, can be reexecuted at a later time.

System Crash : There is a hardware malfunction, or a bug in the database software or the operating system that causes the loss of the content of volatile storage, and bring transaction processing to a halt. The content of nonvolatile storage remains intact, and is not corrupted.

Disk Failure : A disk block loses its content as a result of either a head crash or failure during a data-transfer operation. Copies of the data on other disks, or archival backups on tertiary media, such as tapes are used to recover from the failure.

9.10. STORAGE STRUCTURE

Storage media can be distinguished by their relative speed, capacity and resistance to failure, and are classified as volatile storage nonvolatile storages & stable storage.

(*i*) Volatile Storage : Information residing in volatile storage does not usually survive system crashes. Examples of such storage are main memory and cache memory.

Access to volatile storage is extremely fast, both because of the speed of the memory access itself and because it is possible to access any data item in volatile storage directly.

(*ii*) Non-volatile Storage : Information residing in non volatile storage survives system crashes : Example of such storage are disk and magnetic tapes. Disks are used for on line storage, whereas tapes are used for archival storage. Both, however are subject to failure (for *e.g.*, headcash), which may result in loss of information. At the current state of technology, nonvolatile storages is slower than volatile storages by several orders of magnitude. This is because disk and tape devices are electro mechanical rather than based entirely on chips, as is volatile storage.

In database systems, disks are used for most non-volatile storage. Other non volatile media are normally used only for backup data. Flash storage though nonvolatile, has insufficient capacity for most database systems.

(*iii*) Stable Storage : Information residing in stable storage is never lost (never should be taken with a grain of salt), since theoretically never cannot be guaranteed. For *e.g.*, it is possible although extremely unlike that a black-hole may envelop the earth and permanently destroy all data. Although stable storage is theoretically impossible to obtain, it can be closely approximated by techniques that make data loss extremely unlike.

The distinction among the various storage types are often less clear in practice than in our presentation. Certain systems provide battery backup, so that some main memory can survive system crashes and power failures. Alternative forms of non-volatile storage, such as optical media, provide an even higher degree of reliability than do disks.

9.11. RECOVERY AND ATOMICITY

Let us consider a simplified banking system and transaction T_j that transfers $ 50 from account A to account B, with initial values of A and B being $ 1000 and $ 2000, respectively. Suppose that a system crash has occurred during the execution of T_j, after output (B_A) has taken place, but before (B_B) was executed, where B_A and B_B denote the buffer blocks on which *A* and *B* reside. Since the memory contents were lost, we do not know the fate of the transaction, thus, we could invoke one of two possible recovery procedures :

(*a*) **Reexecute T_j :** This procedure will result in the value of A becoming $ 900, rather than $ 950. Thus, the system enters an in consistent state.

(*b*) **Do not execute T_j :** The current system state has values of $ 950 and $ 2000 for *A* and *B*, respectively. Thus, the system enters an inconsistent state.

In either case, the database, is left in an inconsistent state, and thus this simple recovery scheme does not work. The reason for this difficulty is that we have modified the database without having assurance that the transaction will indeed commit. Our goal is to perform either all or no database modifications made by T_j. However, if T_j performed multiple database modifications, several output operations, may be required, and a failure may occur after some of these modifications have been made, but before all of them are made.

To achieve our goal of atomicity, we must first output information describing the modifications to stable storage, without modifying the database itself. As we shall see, this procedure will allow us to output all the modifications made by a committed transaction, despite failures. We shall assume that transactions are executed serially in other words, only a single transaction is active at a time.

9.12. LOG BASED RECOVERY

To keep track of the database transactions, the DBMS maintains special files called log files or journals that contain information about all updates. The log file contains information like transaction identifier, type of the log record etc.

Since the information contained in the log files are critical for the database recovery, two separate copies are maintained. In the past the log files were stored on magnetic tapes. But nowadays, DBMS are expected to recover from minor failures very quickly and so the log files are stored online on a fast direct access storage devices (DASD).

Since the online log files are also prone to failures, the log files are periodically archived and are stored in off-line storage. These log files are called "**archive log files.**"

Log

The most widely used structure for recording database modifications is the log. The log is a sequence of log records, recording all the update activities in the database. There are several types of log records. An update log describes a single database write. It has these fields:

#Transaction identifier is the unique identifier of the transaction that performed the write operation.

#Data item identifier is the unique identifier of the data item written. Typically, it is the location on disk of the data item.

#Old value is the value of the data item prior to the write.

#New value is the value that the data item will have after the writer.

Other special log records exist to record significant events during transaction processing, such as that start of a transaction and the commit or abort of a transaction, we denote the various type of log records as:

- <T_1 start.> Transaction T_i has started.
- < $T_i X_j$, V_1, V_2 >. Transaction T_i has performed a write on data item X_j, X_j had value V_1 before the write and will have V_2 after the write.
- < T_i commit> Transaction T_i has committed.
- < T_i abort> Transaction T_i has aborted.

It is of two types : (*i*) Deferred D.M.
(*ii*) Immediate D.M.

(*i*) Deferred Database Modification : The deferred modification technique ensures transaction atomicity by recording all database modifications in the log, but deferring the execution of all write operations of a transaction until the transaction partially commits. Recall that a transaction is said to be partially committed once the final action of the transaction has been executed. The version of the deferred modification technique that we describe in this section assumes that transactions are executed serially.

Suppose that these transactions are executed serially, in the order T_0 followed by T_1, and that the values of accounts A, B and C before the execution took place were \$1000, \$2000, and \$700, respectively. The portion of the log containing the relevant information on these two transactions appears in figure. There are various orders in which the actual outputs can take place to both the database system and the log as a result of the execution of T_0 and T_1.

<T_0 start>

< T_0 A,950>

<T_0 B, 2050>

<T_0 commit>

<T_1 start>

<T_1 C, 600>

<T_1 commit>

***(ii)* Immediate Database Modification:** The immediate modification technique allows database modifications to be output to the database while the transaction is still in the active state. Data modifications written by active transactions are called uncommitted modifications. In the event of a crash or a transaction failure, the system must use the old value field of the log records described to restore the modified data items to the value they had prior to the start of the transaction. The undo operation described next, accomplishes this restoration.

Log	Database
< T_0 start >	
< T_0 A, 1000, 950 >	
< T_0 B, 2000, 2050 >	A = 950
< T_0 commit >	B=2050
< T1 start>	
< T_1 C, 700, 600 >	C=600
< T_0 commit >	

This figure shows one possible order in which the actual outputs took place in both the database system and the log as a result of the execution of T_0 and T_1.

9.13. SHADOW PAGING

The shadow paging scheme does not require the use of a log file in a single user environment, though a log is needed in multi-user environment for concurrency control. Shadow paging considers the database to be made up of a number of fixed-size disk pages or disk blocks for recovery purposes.

The shadow paging technique maintain two directories (two for each database page) during the life of a transaction—a current directory and a shadow directory. When the transaction starts, the two directories are the same. The shadow directory is saved to the disk and the current directory is used by the transaction. The shadow directory is never changed and is used to restore the database in case of a failure.

During the transaction execution, the shadow directory is never modified.

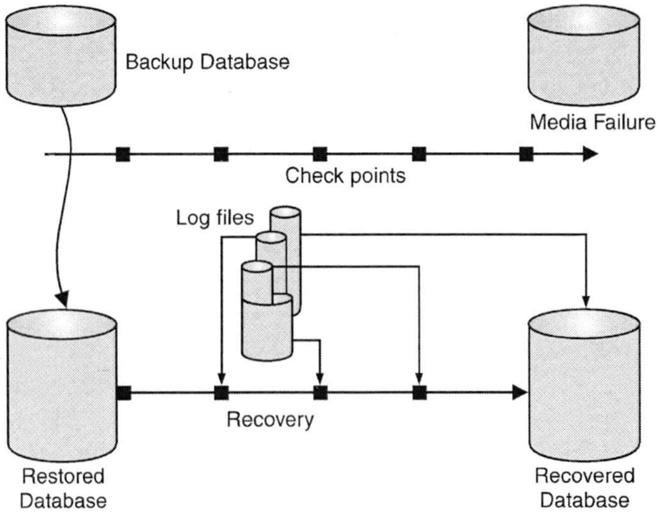

Fig. 9.1. Restoring and recovering a database

9.14. RECOVERY WITH CONCURRENT TRANSACTION

Recovery from concurrent transaction can be done in the various four ways:

1. **Interaction with Concurrency Control:** In this scheme, the recovery scheme depends greatly on the concurrency control scheme that is used. So rollback a failed transaction, we must undo the updates performed by the transaction.
2. **Transaction Rollback:** In this scheme we rollback a failed transaction by using the log. The system scans the log backward, for every log record found in the log the system restores the data item.
3. **Check points:** In this scheme we used check points to reduce the number of log records that the system must scan when it recovers from a crash. In a concurrent transaction-processing system, we require that the checkpoint log record be of the from <checkpoint *L* >, where '*L*' is a list of transactions active at the time of the check point.

   A fuzzy checkpoint is a checkpoint where transactions are allowed to perform updates even while buffer blocks are being written out.
4. **Restart Recovery:** When the system recovers from a crash, it constructs two lists. The undo-list consists of transactions to be undone, and the redo-list consists of transaction to be redone. The system constructs the two lists as follows: Initially, they are both empty. The system scans the log backward, examining each record, until it finds the first <check point> record.

9.15. PROBLEMS IN CONCURRENCY

9.15.1. Lost Updates

Lost updates occur when two or more transactions select the same row and then update the row based on the value originally selected. Each transaction is unaware of other transactions. The last update overwrites updates made by the other transactions, which results in lost data.

For example, two editors make an electronic copy of the same document. Each editor changes the copy independently and then saves the changed copy, thereby overwriting the original document. The editor who saves the changed copy last overwrites changes made by the first editor. This problem could be avoided if the second editor could not make changes until the first editor had finished.

9.15.2. Uncommitted Dependency (Dirty Read)

Uncommitted dependency occurs when a second transaction selects a row that is being updated by another transaction. The second transaction is reading data that has not been committed yet and may be changed by the transaction updating the row.

For example, an editor is making changes to an electronic document. During the changes, a second editor takes a copy of the document that includes all the changes made so far, and distributes the document to the intended audience. The first editor then decides the changes made so far are wrong and removes the edits and saves the document. The distributed document contains edits that no longer exist, and should be treated as if they never existed. This problem could be avoided if no one could read the changed document until the first editor determined that the changes were final.

9.15.3. Inconsistent Analysis (Nonrepeatable Read)

Inconsistent analysis occurs when a second transaction accesses the same row several times and reads different data each time. Inconsistent analysis is similar to uncommitted dependency in that another transaction is changing the data that a second transaction is reading. However, in inconsistent analysis, the data read by the second transaction was committed by the transaction that made the change. Also, inconsistent analysis involves multiple reads (two or more) of the same row and each time the information is changed by another transaction; thus, the term nonrepeatable read.

For example, an editor reads the same document twice, but between each reading, the writer rewrites the document. When the editor reads the document for the second time, it has changed. The original read was not repeatable. This problem could be avoided if the editor could read the document only after the writer has finished writing it.

9.15.4. Phatom Reads

Phantom reads occur when an insert or delete action is performed against a row that belongs to a range of rows being read by a transaction. The transaction's first read of the range of rows shows a row that no longer exists in the second or succeeding read, as a result of a deletion by a different transaction. Similarly, as the result of an insert by a different transaction, the transaction's second or succeeding read shows a row that did not exist in the original read.

For example, an editor makes changes to a document submitted by a writer, but when the changes are incorporated into the master copy of the document by the production department, they find that new unedited material has been added to the document by the author. This problem could be avoided if no one could add new material to the document until the editor and production department finish working with the original document.

SOLVED EXAMPLES

Example 9.1. Show that the two-phase locking protocol ensures conflict serializability and that transactions can be serialized according to their lock points.

Solution : Suppose two-phase locking does not ensure serializability. Then there exists a set of transactions T_0, T_1, ... T_{n-1} which obey 2PL and which produce a non-serializable schedule. A non-serializable schedule implies a cycle in a precedence graph, and we shall show that 2PL cannot produce such cycles. Without loss of generality, assume the following cycle exists in the precedence graph : $T_0 \to T_1 \to T_2 \to \ldots \to T_{n-1} \to T_0$. Let α_i be the time at which T_i obtains its last lock (*i.e.*, T_i's lock point). Then for all transactions such that $T_i \to T_j$, $\alpha_i < \alpha_j$. Then for the cycle we have

$$\alpha_0 < \alpha_1 < \alpha_2 < \ldots < a_{n-1} < \alpha_0$$

Since $\alpha_0 < \alpha_0$ is a contradiction, no such cycle can exist. Hence 2PL cannot produce non-serializable schedules. Because of the property that for all transactions such that $T_i \to T_j$, $\alpha_i < \alpha_j$, the lock point ordering of the transactions is also a topological sort ordering of the precedence graph. Thus transactions can be serialized according to their lock points.

Example 9.2. Consider the following two transactions :

T_{31} : read (A);
read (B);
if A = 0 then B : = B+1;
write (B)

T_{32} : read (B);
read (A);
if B = 0, then A: = A+1;
write (A)

Add lock and unlock instructions to transactions T_{31} and T_{32}, so that they observe the two-phase locking protocol. Can the execution of these transactions result in a deadlock.

Solution : (a) Lock and unlock instructions :

T_{31} : lock-S(*A*)
read (*A*)
lock-X(*B*)
read (*B*)
if *A* = 0
then *B* := *B* + 1
write (*B*)
unlock(*A*)
unlock(*B*)

T_{32} : lock-S(*B*)
read (*B*)
lock-X(*A*)
read (*A*)

if $B = 0$
then $A := A + 1$
write (A)
unlock(B)
unlock(A)

(b) Execution of these transactions can result in deadlock. For example consider the following partial schedule :

T_{31}	T_{32}
lock-S(A)	
	lock-S(B)
	read(B)
read(A)	
lock-X(B)	
	lock-X(A)

The transactions are now deadlocked.

Example 9.3. What benefit does rigorous-two phase locking provide ? How does it compare with other forms of two-phase locking ?

Solution : Rigorous two-phase locking has the advantages of strict 2PL. In addition it has the property that for two conflicting transactions, their commit order is their serializability order. In some systems users might expect this behaviour.

Example 9.4. Show by example that there are schedules possible under the tree protocol that are not possible under the two-phase locking protocol, and vice-versa.

Solution : Consider the tree-structured database graph given below :

Schedule possible under three protocol but not under 2PL :

T_1	T_2
lock(A)	
lock(B)	
unlock(A)	
	lock(A)
lock(C)	
unlock(B)	
	lock(B)
	unlock(A)
	unlock(B)
unlock(C)	

Schedule possible under 2PL but not under tree protocol :

T_1	T_2
lock(*A*)	
	lock(*B*)
lock(*C*)	
	unlock(*B*)
unlock(*A*)	
unlock(*C*)	

Example 9.5. For each of the following protocols, describe aspects of practical applications that would lead you to suggest using the protocol, and aspects that would suggest not using the protocol :

(*a*) Two-phase locking.

(*b*) Two-phase locking with multiple-granularity locking.

(*c*) The tree protocol

(*d*) Timestamp ordering

(*e*) Validation

(*f*) Multiversion timestamp ordering

(*g*) Multiversion two-phase locking.

Solution :

(*a*) **Two-phase locking:** Use for simple applications where a single granularity is acceptable. If there are large read-only transactions, mulliversion protocols would do better. Also, if deadlocks must be avoided at all costs, the tree protocol would be preferable.

(*b*) **Two-phase locking with multiple granularity locking:** Use for an application mix where some applications access individual records and others access whole relations or substantial parts thereof. The drawbacks of 2PL mentioned above also apply to this one.

(*c*) **The tree protocol:** Use if all applications tend to access data items in an order consistent with, a particular partial order. This protocol is free of deadlocks, but transactions will often have to lock unwanted nodes in order to access the desired nodes.

(*d*) **Timestamp ordering:** Use if the application demands a concurrent execution that is equivalent to a particular serial ordering (say, the order of arrival), rather than any serial ordering. But conflicts are handled by roll-back of transactions rather than waiting, and schedules are not recoverable. To make them recoverable, additional overheads and increased response time have to be tolerated. Not suitable if there are long read-only transactions, since they will starve. Deadlocks are absent.

(*e*) **Validation:** If the probability that two concurrently executing transactions conflict is low, this protocol can be used advantageously to get better concurrency and good response times with low overheads. Not suitable under high contention, when a lot of wasted work will be done. Multi version timestamp ordering: Use if timestamp ordering is appropriate but it is desirable for read requests to never wait. Shares the other disadvantages of the timestamp ordering protocol.

(f) **Multi version two-phase locking:** This protocol allows read-only transactions to always commit without ever waiting. Update transactions follow 2PL, thus allowing recoverable schedules with conflicts solved by waiting rather than roll-back. But the problem of deadlocks comes back, though read-only transactions cannot get involved in them. Keeping multiple version.

Summary

- Concurrency control is the management of concurrent transaction execution. DBMS implement concurrency control techniques to ensure serializability and isolation of transaction in order to guard the consistency and integrity of the database.
- Locking is a procedure used to control concurrent access to data. It is one of the most widely used mechanism to ensure serializability.
- The two phase locking protocol ensures serializability. This protocol requires that each transaction issue lock and unlock requests in two phase : Growing Phase and Shrinking Phase.
- Time stamping is a concurrency, control protocol in which the fundamental goal is to order transaction globally in such a way that other transactions get priority in the event of a conflict.
- The modification to the time stamp-ordering protocol is called, "Thomas write Rule".
- In validation-based Protocol techniques we do not updata all the data item directly to the database, rather it keeps local copies of all the updates during transaction execution and as teh transaction reaches it end, a validation phase checks the consistency of the data items due to the updates, and if it is a consistent then all the data from the local copies goes to the database.
- By granularity we mean is the size of the data item being locked. Granularity is considered as the major factor for concurrency control.

Selected Exercises

1. Distinguish between locking and Timestamp protocols for concurrency controls.
2. Explain the 3-phase of validation-based protocol.
3. Explain the Thoma's write Rule.
4. Define the Graph-based protocols.
5. Explain the timestamp ordering protocol.
6. Mention the pros and cons of tree-locking protocol over two-phase locking protocol.
7. What are the various techniques of recovery with concurrent transactions? Explain in details.
8. Explain the multiversion schemes in brief.
9. What are locks? Differentiate shared mode and exclusive mode locks.

Chapter

10 Buffer Management and Query Processing

Inside this Chapter

10.1. LOG-RECORD BUFFERING

So far, we have assumed that every log record is output to stable storage at the time it is created. This assumption imposes a high overhead on system execution for several reasons : Typically, output to stable storage in units of blocks. In most cases, a log record is much smaller than a block. Thus, the output of each log record translates to a much larger output at the physical level.

The cost of outputting a block to stable storage is sufficiently high that it is desirable to output multiple log records at once. To do so, we write log records to a log buffer in main memory, where they stay temporarily until they are output to stable storage. Multiple log records can be gathered in the log buffer and output to stable storage as a single output operation. The order of log records in the stable storage must be exactly the same as the order in which they were written to the log buffer.

As a result of log buffering, a log record may reside in only main memory (volatile storage) for a considerable time before it is output to stable storage. Since such log records are lost if the system crashes, we must impose additional requirements on the recovery techniques to ensure transaction atomicity.

- Transaction *T* enters the commit state after the <*T*, commit> log record has been output to stable storage.
- Before the <*T*, commit> log record can be output to stable storage, all log records pertaining to transaction *T*, must have been output to stable storage.
- Before a block of data in main memory can be output to the database (in non-volatile storage), all log records pertaining to data in that block must have been output to stable storage.

This rule is called the **write-ahead logging (WAL)** rule. (Strictly speaking, the *WAL* rule requires only that the undo information in the log have been output to stable storage. There is no problem resulting from the output of log records *earlier* than necessary. Thus, when the system finds it necessary to output a log record to stable storage, it outputs an entire block of log records, if there are enough log records in main memory to fill a block. If there are insufficient log records to fill the block, all log records in main memory are combined into a partially full block, and are output to stable storage.

Writing the buffered log to disk is sometimes referred to as a **log force.**

Database Buffering

The system stores the database in non-volatile storage (disk) and brings blocks of data into main memory as needed. Since main memory is typically much smaller than the entire database, it may be necessary to overwrite a block B_1 in main memory when mother block B_2 needs to be brought into memory. If B_1 has been modified, B_1 must be output prior to the input of B_2.

The rules for the output of log records limit the freedom of the system to output blocks of data. If the input of block B_2 causes block B_1 to be chosen for output, all log records pertaining to data in B_1 must be output to stable storage before B_1 is output. Thus, the sequence of actions by the system would be :

- Output log records to stable storage until all log records pertaining to block B_1 have been output.
- Output block B_1 to disk.
- Input block B_2 from disk to main memory.

It is important that no writes to the block B_1 be in progress while the system carries out this sequence of actions. We can ensure that there are no writes in progress by using a special means of locking. Before a transaction performs a write on a data item, it must acquire an exclusive lock on the block in which the data item resides. The lock can be released immediately after the update has been performed. Before a block is output, the system obtains an exclusive lock on the block, to ensure that no transaction is updating the block. It releases the lock once the block output has completed. Locks that are held for a short duration are often called **latches.** Latches are treated as distinct from locks used by the concurrency-control system. As a result, they may be released without regard to any locking protocol, such as two-phase locking, required by the concurrency-control system.

To illustrate the need for the write-ahead logging requirement, consider out banking example with transactions T_0 and T_1. Suppose that the state of the log is and that

$$<T_0 \text{ start}>$$

$$<T_0, A, 1000, 950>$$

transaction T_0 issues a read (B). Assume that the block on which B resides is not in main memory and that main memory is full. Suppose that the block on which A resides is chosen to be output to disk. If the system outputs this block to disk and then a crash occurs, the values in the database for accounts A, B and C are 5950, $2000 and $700, respectively. This database state is inconsistent. However, because of the *WAL*

$$<T_0, A, 1000, 950>$$

requirements, the log record must be output to stable storage prior to output of the block on which A resides. The system can use the log record during recovery to bring the database back to a consistent state.

10.2. OVERVIEW OF PHYSICAL STORAGE MEDIA

There are several types of data storage that exist in most computer systems. These storage media are classified by the speed with which data can be accessed, by the cost per unit of data to buy the medium, and by the medium's reliability. Among the physical storage media, some of the typically available are as follows :

(A) Cache: The cache is the fastest and most costly form of storage. Cache memeory is small in size, its use is managed by the computer system hardware.

(B) Main memory: The storage medium used for data that are available to be operated on is main memory. The general purpose machine instructions operate on main memory. Although main memory may contain many megabytes of data, or even hundreds of gigabytes of data in large server systems, it is generally too small for storing the entire database. The contents of main memory are usually lost if a power failure or system crash occurs.

(C) Flash memory: Flash memory differ from main memory in that data survive after power failure. Reading data from flash memory takes less than 100 nanoseconds, which is roughly as fast as reading data from main memory. However, writting data to flash memory is more complicated. Flash memory has found popularity as a replacement for magnetic disks for storing small volumes of data in low-cost computer systems, such as computer systems that are embedded in other devices, in hand-held computers, and in other digital electronic devices such as digital cameras. Flash memory is also used in "USB Keys", which can be plugged into the Universal Serial Bus (USB) slots of computing devices.

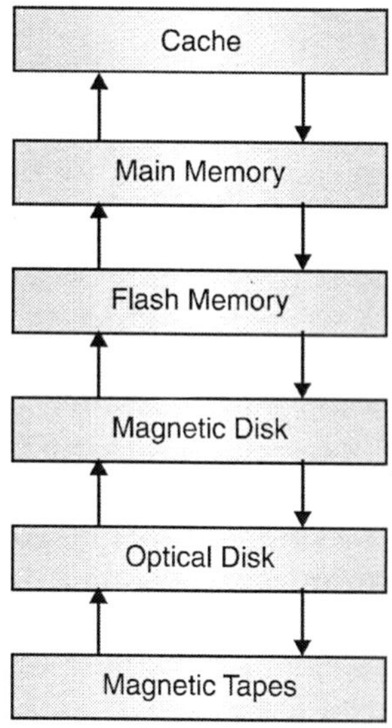

Fig. 10.1.

(D) Magnetic-Disk storage: The primary medium for the long-term and on-line storage of data is the magnetic disk. Usually, the entire databse is stored on magnetic disk. The system must move the data from disk to main memory so that they can be accessed. After the system has performed the designated operations, the data that have been modified must be written to disk. Disk storage survives power failures and system crashes. The size of the magnetic disks currently ranges from a few gigabytes to 400 gigabytes.

(E) Optical Storage: The most popular forms of optical storage are the compact disk (CD), which can hold about 700 megabytes of data and has a playtime of about 80 minutes and the digital video disk (DVD), which can hold 4.7 or 8.5 gigabytes of data per side of the disk. The expansion digital versatile disk is also used in place of digital video disk, since DVDs can hold any digital data, not just video data. Data are stored optically on a disk, and are read by a laser.

(F) Tape Storage: Tape storage is used primarily for backup and archival data. Although magnetic tape is cheaper than disks, access to data is much slower, because the tape must be accessed sequentially from the beginning. For this reason, tape storage is referred to as sequential access storage. In contrast, disk storage is referred to a direct-access storage because it is possible to read data from any location on disk.

Storage Device Hierarchy

The various storage media can be organized in a hierarchy according to their speed and their cost as shown in Fig. 10.1. The higher levels are expensive, but are fast. As we move down the hierarchy, the cost per bit decreases, whereas the access time increases.

Today, almost all active data are stored on disks, except in very rare cases where they are stored on tape or in optical juke boxes.

The fastest storage media such as cache and main memory are referred to as primary storage. The media in the next level in the hierarchy such as magnetic disks are referred to as secondary storage, or online storage. The media in the lowest level in the hierarchy such as magnetic tape and optical-disk are referred to as tertiary storage or off line storage.

In additionm to the speed and cost of the various storage systems, there is also the issue of storage volatility. Volatile storage loses its contents when the power to the device is removed.

10.3. INTRODUCTION TO FILE ORGANIZATION

File organisation can be defined as the method of sorting storing data record in a file and the subsequent implications on the way these records can be accused.

A file organization is way of arranging the various records on the disk. Accessing a record on the file of utmost importance. Records in the file may be organised in several ways and this yields different access mechanise. Since data is transferred between disk storage and main memory, and the basical write of transfer is page or collection of records, it is worth while to arrange the records in a file in such a way that the search for a desired record takes minimum time and effort.

The factors involved in selecting a particular file organisation for uses are :

- Ease of retieval
- Reliability
- Convenient of updates
- Security
- Economy of storage
- Integrity.

Terminology

Field : It is an elementary data item characterised by its size length and types. For example,

Name : a character type of size 15.

Age : a numeric type

Record : It is a collection of related fields that can be treated as a unit from an application point of view. For example, A university could use a student record with the fields, university enrolment no, name, major subjects.

File : Data is organised for storage in files. A file is a collection of similar, related records. It has an identifying name, *e.g.*, 'STUDENT' could be a file consisting of student records for all the pupils in a university.

Index : An index file corresponds to a data file. Its records contain a key and a pointer to that record of the data file which has the same value of the key field.

The data stored in files is accessed by a software which can be divided into two categories.

1. User programs : These are usually written by a programmer to manipulate retrained data in the manner required by the application.

2. File : These deal with the physical movement of data in and out of files. User programs effectively use file operations through appropriate programming language system. The file management system manages the independent files and acts as the software interface between the user programs and the file operations. The some of the basic file operations that can be performed on any kind of files are :

- **Creation of file :** Before creating a file we have to collect the data validate the data and process the data of give it a record like structure. Then we choose the name for the file and open the file on secondary storage device and store the collected data on it.
- **Reading of file :** After creating a file it is needed at any time that we have to read a file. Either we have to any time that we have to read a file. Either we have to read a particular record or the whole file.

  But for both types of read operations firstly, we have to check whether a file which has to be read either exists on the disks or not.
- **Updation of file :** Updation of file means that we have to perform some changes on the contents of file. For updating a file we have to update, then search for the record which has to be changed in the file. If the record is present in the file, then replace that record with the new record, then overwrite the file.
- **Insertion in the file :** Sometimes, it is required to insert a new record in the file. If we have to insert a new record in the end of a file, then the insertion is called appending. If we have to insert a new record after a particular record, then we have to find the position of insertion, these place the new recorded at that position by shifting succeeding records to the right.
- **Deletion from the file :** When some record is not required in the file then we delete that record. For deleting a record we have to search the record for deletion in the file and if it exists delete that record from the file.

10.3.1. Types of File Organisation

These are basically four types of file organisations :

Heap File Organisation

In this organization, any record can be placed anywhere in the file where these is space for the record. These is no ordering of records. Typically, these is a single file for each relation.

Here the records are placed in random order. The records are placed on after another in the file and are totally unordered. This creates of problems in basic file operations such as insertion of a records deletion of a record, modifying the contents of record and searching a desired record. To search a desired record requires the sequential scan of all the records from the beginning until the designed record is found. Therefore, search of the desired record takes a long time. Here file may be logical viewed as :

Record 1
Record 2
....
....
....
Record N

Fig. 10.2. A Heap file

Sequential File Organisation

In this organisation, records are stored in sequential order, based on the value of the search key of each records.

Hashing File Organisation

In this organisation, a hash function is computed on some attribute of each record. The result of the hash function specifies in which block of the file the record should be placed. This organisation is closely related to the indexing structures.

Clustering File Organisation

In this organisation, records of several different relations can be sorted in the same file. Related records of the different relations are stored on the same block, so that one input/output operation fetches related records from all the relations.

10.3.1.1. Sequential File

A sequential file is designed for efficient processing of records in sorted order based on some search key. To permit fast retrieval of records in search-key order, we chain together records by pointers. The pointer in each record points to the next record in search-key order. Further, to minimize the number of block accesses is sequential file processing, we store records physically in search-key order, or as close to search-key order as possible.

The sequential file organisation allows records to be read in sorted order that can be useful for display purposes, as well as for certain busy-processing algorithms. This is the most common structure for large files that are typically processed in their entirety, and its at the heart of the more complex schemes. In this scheme, all the records have the same size and the same fields format, with the fields having fixed size as well. The records are stored in the file according to the content of a field of a scalar type called "Key". The key must identify uniquely a records hence different record have different keys. This organisation is well suited for batch processing of the entire file, without adding or deleting items : this kind of operation can take advantage of the fixed size of records and file; moreover, this organisation is easily stored both on disk and tape. However, adding and deleting records to this kind of file is a tricky process; the logical sequence of records typically matches their physical layout on the media storage, so to ease file navigation, hence adding a record and maintaining the key order requires a reorganisation of the whole file.

Area of Use

Sequential files are most frequently used in commercial batch oriented data processing where these is the concept of a master file to which details are added periodically. For *e.g.*, payroll applications.

The Structure of Sequential Files

In a sequential file, records are stored one after the other on a storage device. All type of external storage devices support a sequential-file organization Magnetic disks and drums provide both direct and sequential access to records and hence support sequential files along with other types of file structures. A sequential file is physically placed on a drum or disk by storing the sequence of records in adjacent locations on a track.

The operations that can be performed on a sequential file may differ slightly depending on the storage device used. A sequential file on disk can be used strictly for input, strictly for output, or for update.

Update means that as records are read, the record most recently real can be rewritten on the same file if so desired.

A sequential file is designed for efficient processing of records. The search for a record in sequential files is based on a key. Every record contains a key by which it can be uniquely identified. These key are called search keys. For the fast retieval of records the search key is used and a pointers chain is developed among the record. Sequential files are also called sorted files. The sequential files are either sorted or nearly sorted. All the records in the sequential file are placed in the order of the search key. This process also reduces the search time as all the records are stored on a criterion based on keys.

Roll number	Name	Father name	Marks
1001	Ram	S.K. Yadav	68
1002	Shyam	B.R. Sagar	75
1003	Yuvraj	P.K. Tripathi	80
1004	Akansha	M.S. Bose	82
1005	Ravi	Rampal	70

Fig. 10.3. Logical view of sequential file

In sequential files, the data is stored and/or retrieved in a logical order, *i.e.*, in a sequence. The records are stored one after another in an ascending or descending order, based on the key field of the records. Sequential files are easy to organise and maintain. The hardware, associated with sequential organisation, is also relatively cheaper, compared to the other file organisations. However, sequential search takes a very long time. If there are N records in the file, on an average, the user will have to go through N/2 records before finding the right one.

Disadvantages

- Updates are not easily accommodated.
- By definition, random access is not possible.
- All records must be structurally identical.

- Continuous area may not be possible because both the primary data file and the transaction file must be looked during merging.

10.3.1.2. Index Sequential Files

When there is need to access records sequentially by some key value and also to access records directly by the same key value, the collection of records may be organised in an effective manner called indexes sequential organisation. The data in the dictionary is stored in sequential manner. However, an index is provided in terms of thems takes. To search for a word are do not search sequential, we access the index that is the appropriate thems tab, locate an appropriate location for the word and these proceed to find the word sequentially.

To implement the concept of indexed sequential file organisation, we consider an approach in which the index part and data fast contain on a separate file. The index file has a tree structure and data files has sequential structure. Since the data file sequential file with a two-level index. Level 1 of the index holds an entry for each three-record section of the main file. The level 2 indexes level 1 in the same way.

When the new records are inserted in the data file, the sequence of records need to be presumed and also the index is accordingly updated.

Two approaches used to implement indexes are static indexes and dynamic indexes.

As the main data file changes due to insertion and detections, the static index contents may change but the structure does not change. In case of dynamic indexing approach. Insertions and deletions in the main data file may lead to changes in the index structure.

The Structure of Indexed Sequential Files

The capability of directly accessing a record based on a key (or unique index can only be achieved if the external storage device used supports this type of access, the file-structure concepts relating to indexed sequential files are best exemplified when considering a magnetic disk as the storage capacity, disks are generally chosen when using indexed sequential files.

There are two types of indexed sequential file organisation the first one is due to IBM, and the second is used by CDC. An IBM indexed sequential file consists of three separate areas : the prime area, the index area, and the overflow are. The prime are is an area into which data records are written when the file is first created. The file is created sequential that is, by written records in the prime are in a sequence dictated by the lexical ordering of the keys of the records.

The second important area of an indexed sequential file, the index area is created. Automatically by the data-management routines in the operating system. A number of index levels may be involved in an indexed sequential file.

The concept of index for a files is similar to the concept of having table of contents in a book. To search an item or topic in a book, the table of contents may be seen first and then the item is searched at the appropriate place. This techniques minimizes the search time as the whole book need not to be searched in sequential order.

Some of the performance criteria for indexing are :

- **Seek Time :** The seek time is defined as the total time to locate a desired record. Access time is minimum using indexed file.

- **Insertion Time :** Insertion time is defined as the time to insert a new record to an appropriate position. Insertion tie includes the time required to find the appropriate location where the new record will be inserted as well as the time required to rebuild the index.

Deletion Time : This is the time required to delete a record. Deletion time include the time required to find the record (access time) and the time taken to rebuild the index after deletion of a record.

In the indexing process, an index structure is used to locate the record in a file. In order to store the index structure additional storage is required. Based on the number of pointers from the index structure, two types of indices are used. These are called sparse index and dense index. In the dense index scheme, an index record in the index structure appears for all key values in the file. The disadvantage of this indexing scheme is that a comparatively larger index structure has to be created and deletion of records become difficult. In the sparse index scheme, index records in the index structure are created for a few.

Key values: To search a record, the search value is compared with the values in the index structure and the pointer is used to identify the block where the desired record lies.

DEPT	ADDRESS
Accounts	120
Marketing	160
Admin	170
Sales	235
Sales	235

Index

Address 170

Sequential access

CODE	NAME	DEPT	SALARY
AD15	P.Ricky	Admin	16,000
AD17	Smith R	Admin	18,000
AD20	C.Aston	Admin	20,000

File

Fig. 10.4. Index sequential access

Essential, indexed sequential technique is a hybrid of sequential and direct file organisation. It provides a combination of access types that are supported by a sequential and a direct file. The indexed file organisation user a separate index file, which contains the key values and the location of the corresponding record. The records are organised in an orderly sequence and the index table is used to access the records without scarehing through entire file. The records may be in random sequence but the index table is tored in sorted sequence on the key values. Since the index table is in sorted sequence, the file management system simply accesses the records in the order of the index values. To access a particular record, a search is made in the index table to determine the address of the first record of the segment in which the data is placed. Then a direct access is made to that address. After that, a sequential search is involved until the desired record is located.

10.3.1.3. Direct File Organisation

Direct files facilitate accessing any record directly or randomly without having to traverse the sequence of the records. These files are also known as random or relative files. Even though only one item can be accessed at a time, but that item may be stored any where in the file. Generally, these files are stored on direct access storage device (DASDs) such as hand disks and CDs. However apart from DASDs, the file itself must have random access facility. Hence, a sequential file stored on a direct access device like CD-ROM will process the data sequentially irrespective of the storage medium, when the records are stored, the program itself generates the address for the records by applying certain techniques to its key field, which is unique for every record. These addresses are stored in a master index, which is loaded into main memory before any database manipulation takes places. It defines the relation ship between the primary key and the corresponding disk address. Since direct search is done in the computer's main memory, it takes less amount of time. Based on the address found in master index, R/W head immediately goes to correct disk position and refiners the desired record. As a result, updation, addition and deletion of record becomes much faster. However this method is expensive than sequential because usually the direct files are sorted on direct access devices which are expensive as compared to sequential storage devices. These over, direct access is less storage efficient because they tend to generate wide gaps in between any two records. In addition, address generation overhead is also involved for accessing records due to addressing functions.

It offers an effective way to organise data when there is a need to access individuals records directly (or random access), a relationship is used to translate the key value into a physical address. This is called the mapping function RR. (Key value Address).

Direct files are stored on DASD (Direct Access Storage Device).

Structure of Direct Files

In a direct (also called random) file, a transformation or mapping is made from the key of record to the address of the storage location at which that record is to reside in the file. One mechanism used for generating this transformation is called a hashing algorithm.

The hashing algorithm used for direct files are very similar to those used for tables, before reading this section. The main conceptual differences are due to the physical characteristics of external storage, which differ from the directly addressable storage characteristics.

10.4. INDEXING

Ordered Indices : Each index structure is associated with a particular search key. An ordered index stores the values of the search keys in stored order and associates with each search key the records that contain it. A file may have several indices, on different search keys. If the file containing the records is sequentially ordered, a **primary index** is an index whose search key also defines the sequential order of the file. Primary indices are also called **clustering indices.** The search key of a primary index is usually the primary key. Indices whose search key specifies an order different from the sequential order of file are called **secondary indices** or **non-clustering** indices.

Primary Index : Files with a primary index on the search key, are called **index-sequential files.** Fig. 10.5, shows a sequential file of *account* records. In the example, the records are stored in search-key order, with *branch-name* used as the search key.

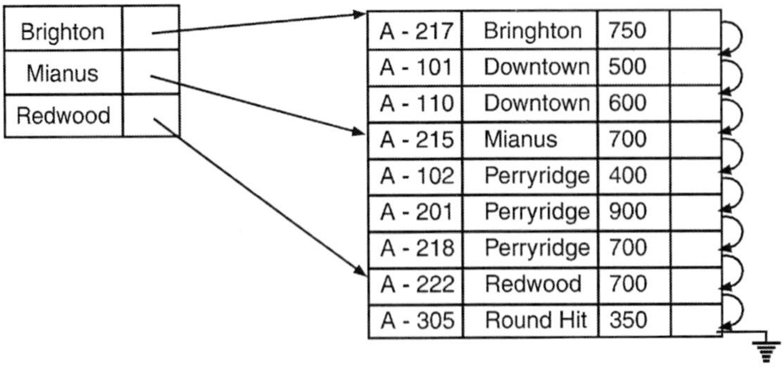

Fig. 10.5. Sequential file for account records.

Dense and Sparse Indices

There are two types of ordered indices that we can use :

Dense index : An index record, appears for every search-key value in the file. In a dense primary index, the index record contains the search-key value and a pointer to the first data record with that search-key value. The rest of the records with the same search key-value would be stored sequentially after the first record.

Sparse index : An index record appears for only some of the search-key values. To locate a record, we find the index entry with the largest search-key value that is less than or equal to the search-key value for which we are looking.

Fig. 10.6 and Fig. 10.7 show dense and sparse indices, respectively, for the *account* file. Suppose that we are looking up records for the 'Perryridge' branch. Using the dense index of Fig. 10.6 we follow the pointer directly to the first 'Perryridge' record. We process this record, and follow the pointer in that record to locate the next record in search-key (branch-name) order. We continue processing records until we encounter a record for a branch other than 'Perryridge'. Since the last entry before the 'Perryridge' is 'Mianus', we follow that pointer. We then read the account file in sequential order until we find the first 'Perryridge' record, and begin processing at that point.

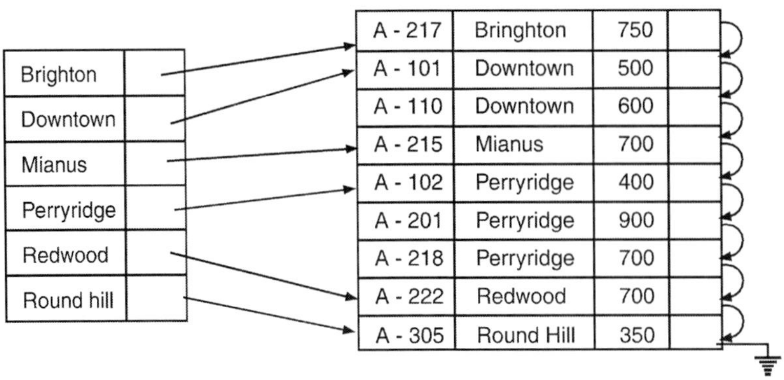

Fig. 10.6. Dense index

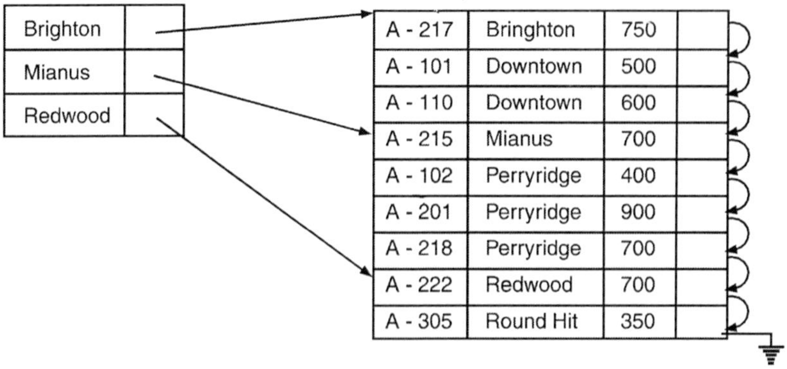

Fig. 10.7. Sparse index

It is generally faster to locate a record if we have a dense index rather than a sparse index. However, sparse indices have advantage over dense indices in that they require less space and they impose less maintenance overhead for insertions and deletions.

Secondary Indices : Secondary indices must be dense, with an index entry for every search-key value, and a pointer to every record in the file. A secondary index on a candidate key looks just like a dense primary index, except that records pointed to by successive values in the index are not stored sequentially. However, secondary indices may have a different structure from primary indices. If the search key of a primary index is not a candidate key, it suffices if, the index points to the first record with a particular value for the search key, since the other records can be fetched by a sequential scan of the file.

In contrast, if the search key of a secondary index is not a candidate key, it is not enough to point to just the first record with each search-key value. The remaining records with the same search-key value could be anywhere in the file, since the records are ordered by search key of the primary index, rather than by the search key of the secondary index. Therefore, a secondary index must contain pointers to all the records.

We can use an extra level of indirection to implement secondary indices on search keys that are not candidate keys. The pointers is such a secondary index do not point directly to the file. Instead, each points to a bucket that contains pointers to the file. Fig. 10.8 shows the structure of a secondary index that uses an extra level of indirection on the *account* file, on search key *balance*.

Advantage : Secondary indices improve the performance of queries that use keys other than the search key of the primary index.

Disadvantage : Secondary indices imposes a significant overhead on modification of the database.

Multilevel Indices : *"Indices with two or more levels are called multilevel indices."*

The problem with sparse index is, the index itself may become too large for efficient processing. Consider a file with 1,00,000 records, with 10 records stored in each block. If we have one index record per block, the index has 10,000 records. Index records are smaller than data records, so let us assume that 100 index records fit on a block. Thus, our index occupies 100 blocks. Such large indices are stored as sequential files on disk.

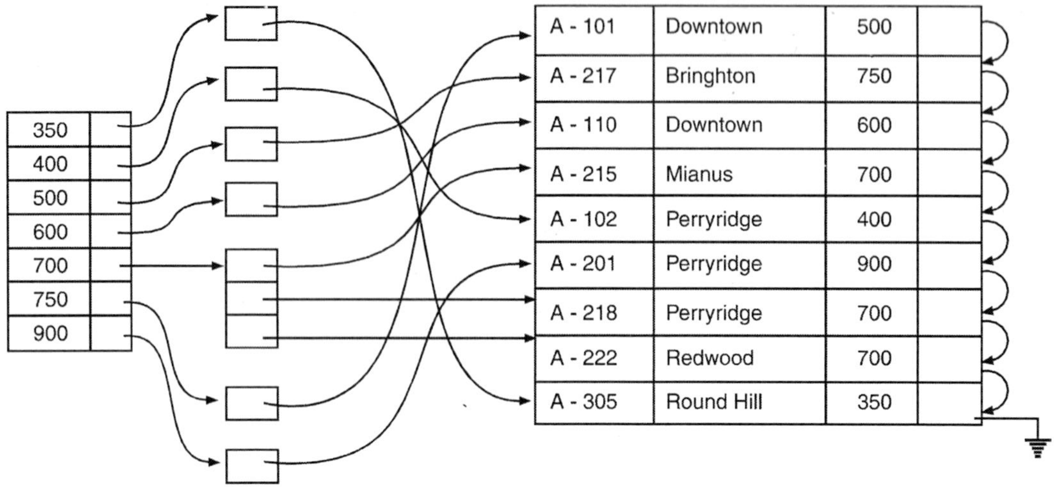

Fig. 10.8. Secondary index on account file, on noncandidate key balance.

If an index is sufficiently small to be kept in main memory, the search time to find an entry is low. However, if the index is so large that it must be kept on disk, a search for an entry requires several disk block reads. If the index occupies *b* blocks, binary search requires as many as $[\log_2(b)]$ blocks to be read. For 100-block index, binary search requires seven block reads. On a disk system where a block read takes 30 ms, the search will take 210 ms, which is long.

To deal with problem, we use two level index structure as in Fig. 10.9. To locate a record, we first use binary search on the outer index to find the record for the largest search-key value less than or equal to the one that we desire. The pointer points to a block of the inner index. We scan this block until we find the record that has the largest search-key value less than or equal to the one that we desire. The pointer in this record points to the block of the file that contains the record for which we are looking.

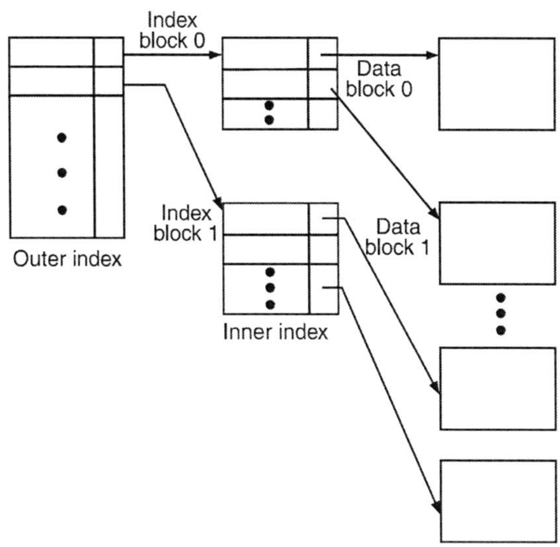

Fig. 10.9. Two-level sparse index

If we assume that the outer index is already in memory, then using the two levels of indexing, we have read only one index block, rather than the seven we read with binary search. If our file is extremely large, even the outer index may grow too large to fit in main memory. In such a case, we can create yet another level of index. Searching for records with a multilevel index requires significantly fewer I/O operations than binary search. A typical dictionary is an example of a multilevel index.

10.5. HASHING TECHNIQUES

One disadvantage of sequential file organization is that we must access an index structure to locate data or must use binary search and that results in more I/O operations. File organization based on the technique of hashing allow us to avoid accessing an index structures. Hashing also provides a way of constructing indices.

There are two types of hashing techniques :

- Static Hashing
- Dynamic Hashing.

10.5.1. Static Hashing

In a hash file organization, we obtain the address of the disk block containing a desired record directly by computing a function on the search key value of the record. In our description of hashing, we shall use the term bucket to denote a unit of storage that can store one or more records. A bucket is typically a disk block, but could be chosen to be smaller or larger than a disk block.

Formally, let K denote the set of all search key values and let B denote the set of all bucket addresses. A hash function h is a function from K to B. Let h denote hash function.

To insert a record with search key K we compute $h(K_1)$ which gives the address of the bucket for that record. Assume that there is space in the bucket to store the record. Then, the record is stored in that bucket.

To perform a lookup on a search key value K_1 we simply compute $h(K_1)$ then search the bucket with that address. Suppose that two search keys, K_5 and K_7, have the same hash value. That is, $h(K_5) = h(K_7)$. If we perform a lookup on K_3, the bucket $h(K_5)$ contains records with search key values K_5 and records with search key values K_7. Thus we have to check the search key value of every record in the bucket to verify that the record is one that we want.

Deletion is equally straightforward. If the search-key value of the records to be deleted is K_7 we compute $h(K_7)$, then search the corresponding bucket for that record, and delete the record from the bucket.

Hash Functions : An ideal hash function distributes the stored keys uniformly across all the buckets, so that every bucket has the same number of records. We want to choose a hash function that assigns search-key values to buckets in such a way that the distribution has these qualities.

- The distribution is *uniform*. That is the hash function assigns each bucket the same number of search-key values from the set of all possible search-key values.
- The distribution is *random*. That is in the average case, each bucket will have nearly the same number or values assigned to it, regardless of the actual distribution of search key values. More precisely, the hash value will not be correlated to any externally visible ordering on the search-key values, such as

alphabetic ordering or ordering by the length of the search keys, the hash function will appear to be random.

Handling of Bucket Overflows

When a record is inserted, the bucket to which it is mapped has space to store the record. If the bucket does not have enough space, a bucket overflow is said to occur. Bucket overflow can occur for several reasons.

- **Insufficient buckets :** The number of buckets which we denote n_B, must be chosen such that $n_B > n_r/f_r$, where n_r denotes the total number of records that will be stored and f_r denotes the number of records that will fit in a bucket. This designation, of course assumes that the total number of records is known when the hash function is chosen.
- **Skew :** Some buckets are assigned more records than others, so a bucket may overflow even when other buckets still have space. This situation is called bucket skew. Skew can occur for two reasons :
  1. Multiple records may have the same search key.
  2. The chosen hash function may result in non-uniform distribution of search keys.

The number of bucket is chosen to be $(n_r/f_r) * (1 + d)$, where d is a fudge factor, typically around 0.2.

We handle bucket overflow by using **overflow buckets.** If a record must be inserted into a bucket b and b is already full, the system provides overflow bucket for b and inserts the record into the overflow bucket. If the overflow bucket is also full, the system provides another overflow bucket, and so on. All the overflow buckets of a given bucket are chained together in a linked list, as in Fig. 10.10. Overflow handling using such a linked list is called overflow chaining.

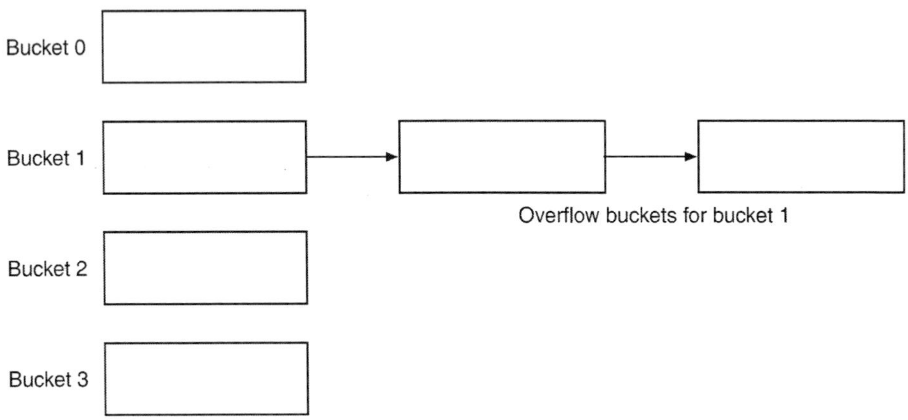

Fig. 10.10. Overflow chaining in a hash structure.

We must change the lookup algorithm slightly to handle overflow chaining. As before the system uses the hash function on the search key to identify a bucket b. The system must examine all the records in bucket b to see whether they match the search key as before. In addition, if bucket b has overflow buckets the system must examine the records in all the overflow buckets also.

The form of hash structure that we have just described is sometimes referred as

closed hashing. Under an alternative approach, called **open hashing,** the set of buckets is fixed, and there are no overflow chains. Instead, if a bucket is full, the system inserts records in some other bucket in the initial set of buckets *b.* One policy is to use the next bucket (in cyclic order) that has space; this policy is called *linear probing.* Other policies, such as computing further has functions, are also used. Open hashing has been used to construct symbol tables for compilers and assemblers, but closed hashing is troublesome. Usually, compilers and assemblers perform only lookup and insertion operations on their symbol tables. However, in a database system, it is important to be able to handle deletion as well as insertion. Thus, open hashing is of only minor importance in database implementation.

Hash Indices

Hashing can be used for index-structure creation. A hash index organizes the search keys, with their associated pointers, into a hash file structure. We construct a has index as follows : apply hash function on a search key to identify a bucket and store the key and its associated pointers in the bucket (or in overflow buckets). Fig. 10.7 shows a secondary hash index on the account file, for the search key account-number. The hash function in the Fig. 10.11 computes the sum of the digits of the account number module 7. The hash index has seven buckets, each of size 2. One of the buckets has three keys mapped to it, so it has an overflow bucket. (Fig. 10.11)

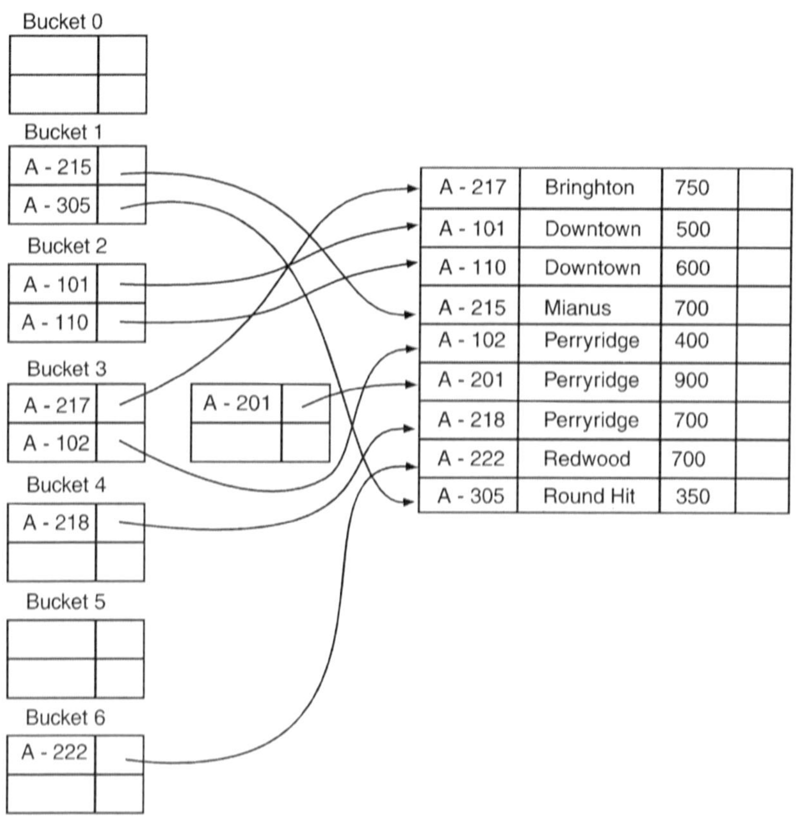

Fig. 10.11. Hash index on search key account-number of account file.

10.5.2. Dynamic Hashing

Most database grow longer over time. If we use static hashing for such database, we have three options :

1. Choose a hash function based on the current file size. This option will result in performance degradation as the database grows.
2. Choose a hash function based on the anticipated size of the file at some point in the future. Although performance degradation is avoided, a significant amount of space may be wasted initially.
3. Periodically reorganize the hash structure in response to file growth. Such a reorganization involves choosing a new hash function, recomputing the hash function on every record in the file and generating new bucket assignments. This reorganization is massive, time-consuming operation.

These problems can be avoided using dynamic hashing. *Dynamic hashing* techniques allow the hash function to be modified dynamically to accommodate the growth or shrinkage of the database. One of the form of dynamic hashing is extendable hashing.

Extendable Hashing

Extendable hashing handles the changes in the database size by splitting and coalescing buckets as the database grows and shrinks. As a result, space efficiency is retained. Moreover, since the reorganization is performed on only one bucket at a time, the resulting performance overhead is acceptably low.

With extendable hashing, we choose hash function *h* with the desirable properties of uniformity and randomness. We do not create a bucket for each hash value. Instead, we create buckets on demand, as records are inserted into the file. We do not use the entire *b* bits of the hash value initially. At any point, we use *i* bits, where 0 < *i* £ *b*. These *i* bits are used as an offset into an additional table of bucket addresses. The value of *i* grows and shrinks with the size of the database.

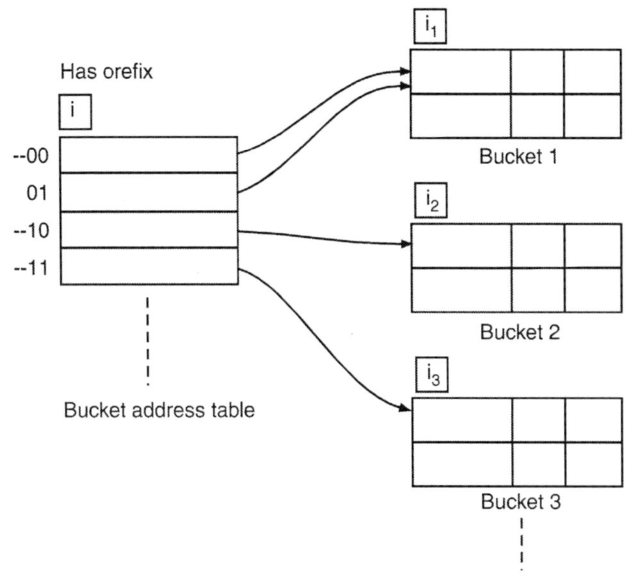

Fig. 10.12. General extendable hash structure

Fig. 10.12 shows a general extendable hash structure. The i appearing above the bucket address table in the Fig. 10.12 indicates that i bits of the hash value $h(K)$ are required to determine the correct bucket for K. This number will change as the file grows.

10.6. B+ TREE INDEX FILES

The B+-Tree index structure is the most widely used of several index structures that maintain their efficiency despite insertion and deletion of data. A B+-tree index takes the form of a balanced tree in to a leaf of the tree is of the same length. Each non leaf node is the tree has between [n/2] and 'n' children where 'n' is fixed for a particular tree.

Structure of a B+-Tree

A B+-tree index is a multilevel index, but it has a structure that differ from that of the multilevel index sequential file. A typical node of B+-tree is as below.

P_1	K_1	P_2	....	P_{n-1}	K_{n-1}	P_n

Fig. 10.13. Typical node of a B+-tree

It contains upto $(n - 1)$ search key values K_1, K_2 ... K_{n-1}, and 'n' pointers P_1, P_2, ... P_n. The search-key values within a node are kept in sorted order thus if $i > j$ the $K_i < K_j$.

An important difference between B+-tree structures and in-memory tree structures. Such as binary tree, each node is small, and has at most two pointers. In a B+-trees each node is large, typically a disk block and a node can have a large number of pointers. Thus B+-trees tend to be fat and short unlike thin and tall binary trees.

Updates on B+-Trees

Insertion and deletion are more complicated than lookup, since it may be necessary to split a node that becomes too large as the result of an insertion, or to combine nods if a node becomes too small. Further where anode is split or a pair of nodes is combined, we must ensure that balance is preserved.

B+-Tree File Organisation

In a B+-tree file organisation, the leaf nodes of the tree store records, instead of strong pointers to records. Since records are usually larger than pointers, the maximum number of records that can be stored in a leaf node is less than the number of pointers in a non-leaf node. However, the leaf node are still required to be at least half full.

Insertion and deletion from a B+-tree file organisation are handled in the same way as insertion and deletion of entries in a B+-tree index.

When a B+-tree is used for file organisation, space utilization is particularly important, since the space occupied by the records is likely to be much more than the space occupied by keys and pointers. We can improve the utilization of space in a B+-tree by involving more sibling nodes in redistribution during splits and merges.

10.7. QUERY PROCESSING

10.7.1. Overview of Query Processing

Query processing refers to the range of activities involved in extracting data from a database. The activities include translation of queries in high-level database languages, into expressions that can be used at the physical level of the life system, a variety of query-optimizing transformations, and actual evaluation of queries.

The steps involved in processing a query appear in Fig. 10.14.

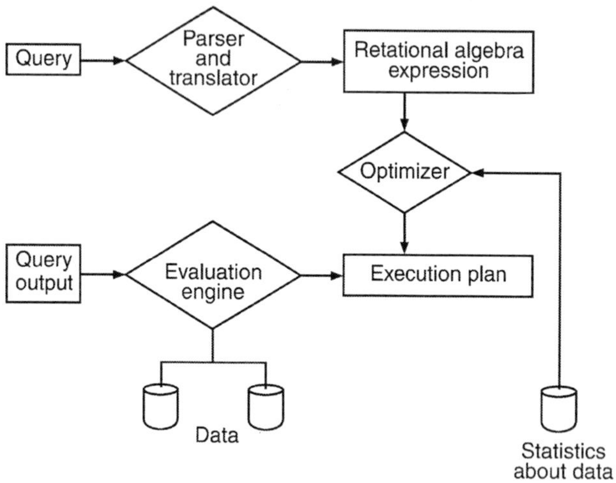

Fig. 10.14. Steps in query processing.

1. Parsing and translation
2. Optimization
3, Evaluation

The first action, the system must take in query processing is to translate a given query into its internal form. This translation process is similar to the work performed by the parser of a compiler. In generating the internal form of the query, the parser checks the syntax of the user's query, verifies that the relation names appearing in the query are names of the relation names appearing in the query are names of the relations in the database and so on. The system construct a parse-tree representation of the query, which it then translates into a relational-algebra expression.

Furthermore, the relational-algebra representation of a query species only partially how to evaluate a query. As an illustration, consider the query :

select balance
from account
where balance < 2500

This query can be translated into either of the following relational-algebra expressions :

- *balance* < 2500 [$P_{balance}$ (*account*)]
- *balance* [$s_{balance < 2500}$ (*account*)]

To implement the preceding selection, we can search every tuple in account to find tuples with balance less than 2500. If a B^+-tree index is available on the attribute *balance*, we can use the index instead to locate the tuples.

To specify fully how to evaluate a query, we need to provide not only the relational-algebra expression, but also to annotate it with instructions specifying how to evaluate each operation.

$\Pi_{balance}$
|
|
$\sigma_{balance < 2500;}$ *use index 1*
|
|
account

Fig. 10.15. A query evaluation plan.

A relational-algebra operation annotated with instructions on how to evaluate it is called an **evaluation primitive.** A sequence of primitive operations that can be used to evaluate a query is a **query-execution plan** or **query-evaluation plan.** Fig. 10.15 illustrates an evaluation plan for our example query, in which a particular index (denoted in the Fig. 10.15 as "index 1") is specified for the selection operation. The query-execution engine takes a query-evaluation plan, executes that plan and returns the answers to the query. The different evaluation plans for a give query can have different costs.

10.7.2. Measures of Query Cost

The cost of query evaluation can be measured in terms of a number of different resources, including disk accesses, *CPU* time to execute a query and in a distributed or parallel database system, the cost of communication.

The response time for a query-evaluation plan (that is, the clock time required to execute the plan) could be used as a good measure of the cost of the plan.

In large database systems, however, disk accesses are usually the most important cost, since disk accesses are slow compared to in-memory operations. Most people consider the disk-access cost a reasonable measure of the cost of a query-evaluation plan.

The number of block transfers from disk is also used as a measure of the actual cost. We also need to distinguish between reads and writes of blocks, since it takes more time to write a block to disk than to read a block from disk. For more accurate measure find out :

1. The number of seek operations performed,
2. The number of blocks read,
3. The number of blocks written,

and then add up these numbers after multiplying them by the average seek time, average transfer time, average transfer time for reading a block, and average transfer time for writing a block, respectively.

10.7.3. Selection Operation

Consider a selection operation on a relation whose tuples are stored together in one file. Two scan algorithms to implement the selection operation are given below.

Basic Algorithms

***A1* (linear search) :** In a linear search, the system scans each file block and tests all records to see whether they satisfy the selection condition. For a selection on a key attribute, the system can terminate the scan if the required records is found, without looking at the other records of the relation.

The cost of linear search, in terms of number of *I/O* operations, is b_r, where b_r denotes the number of blocks in the files. Selection on key attributes have an average cost of $b_r/2$ but still have a worst-case cost of b_r.

Advantage

The linear search algorithms can be applied to any file.

Disadvantage

It is slower.

***A2* (binary search) :** If the file is ordered on an attribute, and the selection condition is an equality comparison on the attribute, we can use a binary search to locate records that satisfy the selection. The system performs the binary search on the blocks of the file.

The number of blocks that need to be examined to find a block containing the required records is $(\log_2(b_r))$, where b_r denotes the number of blocks in the file. If the selection is on a non-key attribute, more than one block may contain required records and the cost of reading the extra blocks has to be added to the cost estimate.

Selections Using Indices

Search algorithms that use an index are referred to as index scans. Search algorithms that use an index are :

- ***A3* (Primary index, equality on key) :** For an equality comparison on a key attribute with a primary index, we can use the index to retrieve a single record that satisfies the corresponding equality condition.

  If a B^+-tree is used, the cost of the operation in terms of *I/O* operations, is equal to the height of the tree plus one *I/O* to fetch the record.
- ***A4* (Primary index, equality on non-key) :** We can retrieve multiple records by using a primary index when the selection condition specifies an equality comparison on a nonkey attribute, *A*. The cost of the operation is proportional to the height of the tree, plus the number of blocks containing records with the specified search key.
- ***A5* (Secondary index, equality) :** Selection specifying an equality condition can use a secondary index. This strategy can retrieve a single record if the equality condition is on a key; multiple records may get retrieved if the indexing field is not a key.

  In the first case, only one record is retrieved and the cost is equal to the height of the tree plus one *I/O* operation to fetch the record. In the second case, each record may be resident on a different block, which may result in one *I/O* operation per retrieved record.

Selections Involving Comparisons

Consider a selection of the form $a_{A < v}(r)$. We can implement the selection either by using a linear or binary search or by using indices in one of the following ways :

- ***A6* (Primary index, comparison) :** A primary ordered index (for example, a primary B^+-tree index) can be used when the selection condition is a comparison. For comparison conditions of the form $A > v$ or $A \geq v$, a primary index on A can be used to direct the retrieval of tuples, as follows. For $A \geq v$, we look up the value v in the index to find the first tuple in the file that has a value of $A = v$. A file scan starting from that tuple up to the end of the file returns all tuples that satisfy the condition. For $A > v$, the file scan starts with the first tuple such that $A > v$.

  For comparisons of the form $A < v$ or A £ v, an index lookup is not required. For $A < v$, we use a simple file scan starting from the beginning of the file and continuing up to the first tuple with attribute $A = v$. the case A £ v is similar, except the scan continues up to the first tuple with attribute $A > v$. In either case index is not useful.
- ***A7* (Secondary index, comparison) :** We can use a secondary ordered index to guide retrieval for comparison conditions involving <, £, > or ^{3}. The lowest-level index blocks are scanned, either from the smallest value up to v (for v and £), or from v up to the maximum value (for > and ^{3}).

Implementation of Complex Selections

We now consider more complex selection predicates.

- **Conjunction :** A conjunctive selection is a selection of the form $\sigma_{\theta_1 \wedge \theta_2 \wedge ..... \wedge \theta_n}(r)$.
- **Disjunction :** A disjunctive selection is a selection of the form $\sigma_{\theta_1 \vee \theta_2 \vee ..... \vee \theta_n}(r)$.
- **Negation :** The result of selection s – q (r) is the set of tuples of r for which the condition q evaluates to false.

We can implement a selection operation involving either a conjunction or a disjunction of simple conditions by using one of the following algorithms :

- **A8 (Conjunctive selection using one index) :** We first determine whether an access path is available for an attribute in one of the simple conditions. If one is, one of the selection algorithms *A*2 through *A*7 can retrieve records satisfying that condition.
- **A9 (Conjunctive selection using composite index) :** An appropriate composite index (that is, an index on multiple attribute) may be available for some conjunctive selections. If the selection species an equality condition on two or more attributes and a composite index exists on these combined attribute fields, then the index can be searched directly. The type of index determines which of algorithms *A*3, *A*4, or *A*5 will be used.
- **A10 (Conjunctive selection by intersection of identifiers) :** Another alternative for implementing conjunctive selection operations involves the use of record pointers or record identifiers. This algorithm requires indices with record pointers, on the fields involved in the individual conditions. The algorithm scans each index for pointers to tuples that satisfy an individual condition. The intersection of all the retrieved pointers is the set of pointers to tuples that satisfy the conjunctive condition. The algorithm then uses the pointers to retrieve the actual records.

  The cost of algorithm *A*10 is the sum of the costs of the individual index scans, plus the cost of retrieving the records in the intersection of the retrieved lists of pointers. This cost can be reduced by sorting the list of pointers and retrieving records in the sorted order. Thereby,

1. All pointers to records in a block come together, hence all selected records in the block can be retrieved using a single *I/O* operation and
2. Blocks are read in sorted order, minimizing disk arm movement.

- ***A*11 (Disjunctive selection by union of identifiers) :** It access path are available on all the conditions of a disjunctive selection, each index is scanned for pointers to tuples that satisfy the individual condition. The union of all the retrieved pointers yields the set of pointers to all tuples that satisfy the disjunctive condition. We then use the pointers to retrieve the actual records. However, if even one of the conditions does not have an access path, we will have to perform a linear scan of the relation to find tuples that satisfy the condition.

10.7.4. Sorting

The process of arranging data either in ascending or descending order is called sorting. Sorting of data plays an important role in database systems for two reasons. First, *SQL* queries can specify that the output be sorted. Second and equally important for query processing, several of the relational operations, such as joins, can be implemented efficiently if the input relations are first sorted.

We can sort a relation by building and index on the sort key and then using that index to read the relation in sorted order. However, such a process orders the relation only logically, through an index, rather than physically. Hence, the reading of tuples in the sorted order may lead to a disk access (disk seek plus block transfer) for each record, which can be very expensive, since the number of records can be much larger than the number of blocks. For this reason, it may be desirable to order the records physically.

The problem of sorting has been studied extensively, both for relations that fit entirely in main memory and for relations that are bigger than memory. In the first case, standard sorting techniques such as quick-sort can be used. Sorting of relations that do not fit in memory is called external sorting. The most commonly used technique for external sorting is the external sort-merge algorithm. The output of the merge stage is the sorted relation. The output file is buffered to reduce the number of disk write operations.

10.7.5. Join Operation

In this section, we study several algorithms for computing the join of relations and we analyze their respective costs.

Consider the expression :

$$depositor \bowtie customer$$

We assume the following information about the two relations :

Number of records of *customer* : $n_{\text{customer}} = 10{,}000$

Number of blocks of *customer* : $b_{\text{customer}} = 400$

Number of records of *depositor* : $n_{\text{depositor}} = 5000$

Number of blocks of *depositor* : $b_{\text{depositor}} = 100$

Nested-Loop Join

Fig. 10.16 shows a simple algorithm to compute the theta join, of two relations *r* and s. This algorithm is called the nested-join algorithm, since it basically consists of a pair of nested for loops. Relation *r* is called the **outer relation** and relation *s* is called **inner relation** of the join. The algorithm uses the notation $t_r \cdot t_s$, where t_r and t_s are tuples $t_r \cdot t_s$ denotes the tuple constructed by concatenating the attribute values of tuples t_r and t_s. The nested-loop algorithm requires no indices. It is expensive, since it examines every pair of tuples in the two relation. Consider the cost of the nested-loop join algorithm. The number of pairs of tuples to be considered is $n_r * n_s$, where n_r denotes the number of tuples in r_t and n_s denotes the number of tuples in s. For each record in r_t we have to perform a complete scan on s. In the worst case, the buffer can hold only one block of each relation and a total of $n_r * b_s + b_r$ block accesses would be required, where b_r and b_s denote the number of blocks containing tuples of *r* and *s* respectively.

Now consider the natural join of *depositor* and *customer*. Assume that *depositor* is the outer relation and *customer* is the relation in the join. We will have to examine 5000 * 10000 = 50 * 10^6 pairs of tuples. In the worst case, the number of block accesses is 5000 * 400 + 100 = 2,000,100. In the best-case scenario, however, we can read both relations only once and perform the computation. This computation requires at most 100 + 400 = 500 block accesses.

```
for each tuple t, in r do begin
      for each tuple ts in s do begin
            test pair (tr, ts) to see if they satisfy the join condition q
                  if they do, add tr, ts to the result
      end
end
```

Fig. 10.16. Nested-loop join

Block Nested-Loop Join

Fig. 10.17 shows block nested-loop join, which is a variant of the nested-loop join where every block of the inner relation is paired with every block of the outer relation. Within each pair of block, every tuple in one block is paired with every tuple in the other block, to generate all pairs of tuples. As before, all pairs of tuples that satisfy the join condition are added to the result.

```
for each block B, of r do begin
    for each block Bs of s do begin
        for each tuple tr in Br do begin
            for each tuple ts in Bs do begin
                test pair (tr, ts) to see if they satisfy the join condition
                if they do, add tr . ts to the result
            end
        end
    end
end
```

Fig. 10.17. Block nested-loop join

The primary difference in cost between the block nested-loop join and the basic nested-loop join is that, in the worst case, each block in the inner relation *s* is read only once for each block in the outer relation, instead of once for each tuple in the outer relation. Thus, in the worst case, there will be a total of $b_r * b_s + b_r$ block accesses, where b_r and b_s denote the number of blocks containing records of *r* and *s* respectively.

In the worst case, we have to read each block of customer once for each block of depositor. Thus, in the worst case, a total of 100 * 400 + 100 = 40,100 block accesses are required. The number of block accesses in the best case remains the same namely, 100 + 400 = 500.

The performance of the nested-loop and block nested-loop procedures can be further improved :

- If the join attributes in a natural join or an equi-join form a key on the inner relation, then for each outer relation tuple the inner loop can terminate as soon as the first match is found.
- If memory has *M* blocks, we read in *M* – 2 blocks of the outer relation at a time and when we read each block of the inner relation we join it with all the *M* – 2 blocks of the outer relation. This change reduces the number of scans of the inner relation from b_r to $[b_r/(M - 2)]$ where b_r is the number of blocks of the outer relation. The total cost is then $[b_r (M - 2)] * b_s + b_r$.
- Reuse the blocks stored in buffer, which reduces the number of disk accesses needed.
- If an index is available on the inner loop's join attribute, we can replace the scans with more efficient index lookups.

Indexed Nested-Loop Join

In a nested-loop join (Fig. 10.18), if an index is available on the inner loop's join attribute, index lookups can replace file scans. For each tuple t_r in the outer relation *r*, the index is used to look up tuples in *s* will satisfy the join condition with tuple t_r.

This join method is called an **indexed nested-loop join;** it can be used with existing indices. For example, consider depositor ⋈ customer. Suppose that we have a depositor tuple with customer-name "John". Then the relevant tuples in *s* are those that satisfy the selection "customer-name = John."

The cost of an indexed nested-loop join can be computed as follows : For each tuple in the outer relation *r* a lookup is performed on the index for *s*, and the relevant tuples are retrieved. Then, b_r disk accesses are needed to read relation *r*, where b_r denoted the number of block containing records of *r*. For each tuple in *r*, we perform an index lookup on *s*. Then, the cost of the join can be computed as $b_r + n_r * c$, where *b*, is the number of records in relation *r* and *c* is the cost of a single selection on *s* using the join condition.

For example, consider an indexed nested-loop join of *depositor* ⋈ *customer*, with depositor as the outer relation. Suppose also that customer has a primary B^+-tree index on the join attribute customer name, which contains 20 entries on an average in each index node. Since customer has 10,000 tuples, the height of the tree is 4, and one more access in needed to find the actual data. Since $n_{depositor}$ is 5000, the total cost is 100 + 5000 * 5 = 25,100 disk accesses. This cost is lower than the 40,100 accesses needed for a block nested-loop join.

Merge Join

The merge join algorithm (also called the sort-merge join algorithm) can be used to compute natural join and equi-joins. Let *r*(*R*) and *s*(*S*) be the relations whose natural join is to be computed and let $R \cap S$ denote their common attributes.

```
pr : = address of first tuple of r;
ps : = address of first tuple of s;
while (ps ≠ null and pr ≠ null) do
begin
    t_s = tuple to which ps points;
    s_s : = {t_s};
    set ps to point to next tuple of s;
    done : = false;
    while (not done and ps ≠ null) do
    begin
        t_s : = tuple to which ps points;
        if (t_s, [JoinAttrs] = t_s[JoinAttrs])
        then begin
            s_s : = s_s ∪ {t'_s};
            set ps to point to next tuple of s;
            end
        else
            done : = true;
        end
        t_r : = tuple to which pr points;
        while (pr ull and t_r[JoinAttrs] < t_s[JoinAttrs]) do
        begin
            set rpr to point to next tuple of r;
            t_r : = tuple to which pr points;
        end
        while (pr null and t_r[JoinAttrs] = t_s[JoinAttrs]) do
        begin
            for each t_s in s_s do
            begin
                add t_s ⋈ t_r to result;
            end
        set pr to point to next tuple of r;
        t_r : = tuple to which pr points;
        end
    end
```

Fig. 10.18. Merge join

Fig. 10.13 shows the merge join algorithm. In the algorithm, *JoinAttrs* refers to the attributes in $R \cap S$ and $t_r \bowtie t_s$, where t_r and t_s are tuples that have the same values for *JoinAttrs*, denotes the concentration of the attributes of the tuples, followed by projecting out repeated attributes. The merge join algorithm associates one pointer with each relation. These pointers point initially to the first tuple of the respective relation. As the algorithm proceeds, the pointer moves through the relation. A group of tuples of one relation with the same value on the join attributes is read into S_s. The algorithm requires that every set of tuples S_s fit in memory. If the relations are in sorted order, the number of block accesses is equal to the sum of the number of blocks in both files, $b_r + b_s$.

Hash Join

The hash join algorithm can be used to implement natural joins and equi-joins. In the hash join algorithm, a hash function h is used to partition tuples of both relations. The basic idea is to partition the tuples of each of the relations into sets that have the same hash value on the join attributes.

We assume that:

- h is a hash function mapping *JoinAttrs* values to $\{0, 1, ..., n_b\}$, where *JoinAttrs* denotes the common attributes of r and s used in natural join.
- H_r, denote partitions of r tuples, each initially empty. Each tuple $t_r \in r$ is put in partition, where $i = h(t_r[JoinAttrs])$.
- H_s, denote partitions of s tuples, each initially empty. Each tuple $t_s \in s$ is put in partition, where $t = h(t_s[JoinAttrs])$.

The idea behind the hash join algorithm is this : Suppose that a r tuple and a s tuple satisfy the join condition; then, they will have the same value for the join attributes. If that value is hashed to some value i, the r tuple has to be in H_{ri} and the s tuple in H_{st}. Therefore, r tuples in H_{rs} need only to be compared with s tuples in H_{st}; they do not need to be compared with s tuples in any other partition.

For example, if d is a tuple in *depositor*, c is a tuple in *customer*, and h is a hash function on the *customer-name* attribute of the tuples, then d and c must be tested only if $h(c) = h(d)$. If $h(c) \neq h(d)$, then c and d must have different values for *customer-name*. However, if $h(c) = h(d)$, we must test c and d to see whether the values in their join attributes are the same, since it is possible that c and d have different *customer-names* that have the same hash value Fig. 10.19 shows the details of the **hash join** algorithm to compute the natural join of relations r and s.

```
/* Partitions */
for each tuple t_s in s do begin
    i : = h (t_s [JoinAttrs]);
    H_si : = H_si ∪ {t_s};
end
/* Partition r */
for each tuple t_r in r do begin
    i : = h (t_r [JoinAttrs]);
    H_ri : = H_ri ∪ {t_r};
end
/* Perform join on each partition */
for i = 0 to n_n do begin
    read H_si and build an in-memory hash index on it
    for each tuple t_r in H_ri do begin
    probe the hash index on H_st to locate all tuples t_s
        such that t_s [JoinAttrs] = t_r [JoinAttrs]
        for each matching tuple t_s in H_si do begin
            add t_r ⋈ t_s to the result
        end
    end
end
```

Fig. 10.19. Hash join

10.7.6. Evaluation of Expressions

One way to evaluate an expression is simply to evaluate one operation at a time, in an appropriate order. The result of each evaluation is **materialized** in a temporary relation for subsequent use. A disadvantage to this approach is the need to construct the temporary relations, which must be written to disk. An alternative approach is to evaluate several operations simultaneously in a **pipeline,** with the result of one operation passed on to the next, without the need to store a temporary relation. Following section describes both the approaches used for evaluation of expressions.

1. Materialization

Consider the expression :

$$P_{customer\text{-}name} (s_{balance < 2500} (account) \bowtie customer)$$

The pictorial representation of above expression is given below :

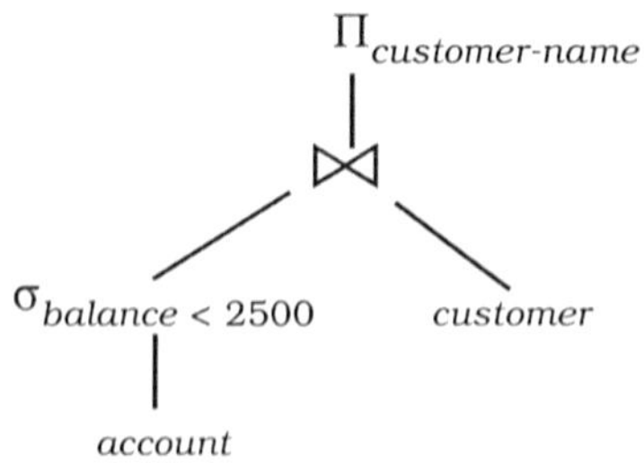

Fig. 10.20. Pictorial representation of an expression.

If we apply the materialization approach, we start from the lowest-level operations in the expression. In our example, there is only one such operation; the selection operation an *account.* The inputs to the lowest-level operations are relations in the database. We execute these operations by the algorithms that we studied earlier and we store the results in temporary relations. We can use these temporary relations to execute the operations at the next level up in the database. In our example, the inputs to the join are the customer relation and the temporary relation created by selection on account. The join can now be evaluated, creating another temporary relation.

By repeating the process, we will eventually evaluate the operation at the roof of the tree, giving the final result of the expression. In our example, we get the final result by executing projection operation at the roof of the tree, using as input the temporary relation created by the join.

Evaluation as just described is called **materialization evaluation,** since the result of each intermediate operations are created (materialized) and then are used for evaluation of the next-level operations.

The cost of evaluating an expression is the addition of the cost of all operations, as well as the cost of writing the intermediate result to disk.

2. Pipelining

We can improve query-evaluation efficiency by reducing the number oft temporary files that are produced. We achieve this reduction by combing several relational operations. Into a *pipeline* of operation, in which the result of one operation are passed along to next operation in the pipeline. Evaluation as just described is called ***pipelined evaluation*** Combing operation into a pipeline eliminates the cost of reading and writing temporary relation.

For example, consider the expression [$P_{a1,\ a2}$ ($r \bowtie s$)] When the join operation generates a tuple of its result it passed that tuple immediately to the project operation for processing. By combining the join and the projection. We avoid creating the intermediate result and instead the final result directly.

Implementation of Pipelining

Each operation in the pipeline is modeled as a separate process or thread within the system, which takes a stream of tuples from its pipelined inputs and generates stream of tuples for its output. For each pair of adjacent operations in the pipeline, the system creates a buffer to hold tuples being passed from one operation to the next.

In the example of Fig. 10.20, all the operations can be placed in pipeline, which passes the result of the selection to the join as they are generated. In turn, it passes the result of join to the projection as they are generated. In turn, it is passes are low, since result of an operation are not stored for long.

Pipelines can be execited in either of two ways :

1. Demand Driven

In a demand driven pipeline, the system makes repeated requests for tuples from the operation at the top of the pipeline. Each time that an operation receives a request for tuples, it computes the next tuple (or tuples) to be returned and then returns that tuple. If the inputs of the operation are not pipelined, the next tuple(s) to be returned can be computed from the input relations. If it has some pipelined inputs, the operation also makes requests for tuples from its pipelined inputs. Using the tuples received from its pipelined inputs, the operation computes tuples for its output and passes them up to parent.

2. Producer Driven

In a **producer driven pipeline,** operations do not wait for requests to produce tuples, but instead generate the tuples **eagerly.** Each operation at the bottom of a pipeline continually generates output tuples and puts them in its output buffer, until the buffer is full. An operation at any other level of a pipeline generates output tuples when it gets input tuples from lower down in the pipeline, until its output buffer is full. Once the operation uses a tuple from a pipelined input, it removes the tuple from its input buffer. In either case, once the output buffer is full, the operation waits until its parent operation removes tuples from the buffer, so that the buffer has space for more tuples. At this point, the operation generates more tuples, until the buffer is full again. The operation repeats this process until all the output tuples have been generated.

10.7.7. Other Operations

Other relational operations and extended relational operations such as duplicate elimination, projection, set operations, outer join and aggregation can be implemented.

Duplicate Elimination : We can implement duplicate elimination easily by sorting. Identical tuples will appear adjacent to each other during sorting and all but one copy can be removed. With external sort-merge, duplicates found while a run is being created can be removed before the run is written to disk, thereby reducing the number of block transfers. The remaining duplicates can be eliminated during merging and the final sorted run has no duplicates. The worst case cost estimate for duplicate elimination is the same as the worst case cost estimate for sorting of the relation.

We can also implement duplicate elimination by hashing, as in the hash-join algorithms, first the relation is partitioned on the basis of a hash function on the whole tuple, then each partition is read in, and an in-memory hash index is constructed.

Projection : We can implement projection easily by performing projection on each tuple, which gives a relation that could have duplicate records and then removing duplicate records. If the attributes in the projection list include a key of the relation, no duplicates will exist, hence, duplicate elimination is not required. Generalized projection can be implemented in the same way as projection.

Set Operations : We can implement the union, intersection and set-differences operations by first sorting both relations and then scanning once through each of the sorted relations to produce the results. In $r \cup s$, when a concurrent scan of both relations reveals the same tuple in both files, only one of the tuples is retained. The result of $r \cap s$, will contain only those tuples that appear in both relations. We implement set difference, $r\text{-}s$, similarly by retaining tuples in r, only if they are absent in s.

For all these operations, only one scan of the two input relation is required, so the cost is $b_r + b_s$ block transfers if the relations are stored in the same order. Assuming a worst case of one block buffer for each relation, a total of $b_r + b_s$ disk seeks would be required in addition to $b_r + b_s$ block transfers. The number of seeks can be reduced by allocating extra buffer blocks.

If the relations are not sorted initially, the cost of sorting has to be included. Any sort order can be used in-evaluation of set operations, provided that both inputs have that same sort order.

10.8. MATERIALIZED VIEWS

When a view is defined, normally the database stores only the query defining the view. In contrast, a materialized view is a view whose contents are computed and stored. Materialization views constitute redundant data, in that thier contents can be inferred from the view definition and the rest of the database contents. However, it is much cheaper in many cases to read the contents of a materialized view that to compute the contents of the view by executing the query defining the view.

Materialized views are important for improving performance in some applications.

View Maintenance

A problem with materialized view is that they must be kept uo-to-date when the data used in the view definition changes. The task of keeping a materialized view up-to-date with the underlying data is known as **View Maintenance.**

Views can be maintained by manually written code. Another option for maintaining materialized views is to define triggers on insert, delete and updata of each relation in the view definition. The triggers must modify the contents of the materialized view, to take into account the change that caused the trigger to fire. A simplistic way of doing so is to completely recompute the materialized view on every tuple.

A better option is to modify only the affected parts of the materialized view, which is known as "**incremental view maintenance**".

Most database systems perform immediate view maintenance that is, incremental view maintenance is performed as soon as an update occurs, as part of the updating transaction. Some database systems also support deferred view maintenance where view maintenance is deferred to a later time, for e.g., updates may be collected throughout a day, and materialized views may be updated at night. This approach reduces the overhead on update transactions.

10.9. DATABASE TUNING

Database tuning describes a group of activities used to optimize and homogenize the performance of a database. It usually overlaps with query tuning, but refers to design of the database files, selection of the database management system (DBMS) application, and configuration of the database's environment (operating system, CPU, etc.)

Database tuning aims to maximize use of system resources to perform work as efficiently and rapidly as possible. Most systems are designed to manage their use of system resources, but there is still much room to improve their efficiency by customizing their settings and configuration for the database and the DBMS.

DBMS tunings refers to tuning of the DBMS and the configuration of the money and processing resources of the computer running the DBMS. This is typically done through configuring the DBMS, but the resources involved are shared with the host system.

Tuning the DBMS can involve setting the recovery interval (time needed to restore the state of data to a particular point in time), assigning parallelism (the breaking up of work from a single query into tasks assigned to different processing resources), and network protocols used to communicate with database consumers.

10.10. CONCEPTS OF OODBMS

An **object database** (also **object-oriented database management system**) is a database management system in which information is represented in the form of objects as used in object-oriented programming. Object databases are different from relational databases and belongs together to the broader database management system.

Object databases have been considered since the early 1980s and 1990s. Object databases main usage is in object oriented areas.

When database capabilities are combined with object-oriented programming language capabilities, the result is an object-oriented database management system (OODBMS). OODBMS allow object-oriented programmers to develop the product, store them as objects, and replicate or modify existing objects to make new objects within the OODBMS. Because the database is integrated with the programming language, the programmer can maintain consistency within one environment, in that both the OODBMS and the programming language will use the same model of representation. Relational DBMS projects, by way of contrast, maintain a clearer division between the database model and the application.

As the usage of web-based technology increases with the implementation of Intranets and extranets, companies have a vested interest in OODBMS to display their complex data. Using a DBMS that has been specifically designed to store data as objects gives an advantage to those companies that are geared towards multimedia presentation or organizations that utilize computer-aided design (CAD).

Some object-oriented databases are designed to work well with object-oriented programming languages such as Delphi, Ruby, Python, Perl, Java, C#, Visual Basic .NET, C++, Objective-C and Smalltalk; others have their own programming languages. OODBMSs use exactly the same model as object-oriented programming languages.

Object database management systems grew out of research during the early to mid-1970s into having intrinsic database management support for graph-structured objects. The term "object-oriented database system" first appeared around 1985.

10.11. DISTRIBUTED DATABASE

A distributed database system consists of a collection of sites, connected together by the use of the some kind of communication network. Each site is a database system site in its own right and the sites have agreed to work together, so that a user at any site can access anywhere in the network as if the data were all stored at the user's local site. In the other words, each site has its own local database, its own local users, its own local DBMS and transaction management software and own local data communication manager. The distributed database system can thus be regarded as a kind of partnership among the individual local DBMS at the individual local sites. A distributed database is fragmented into smaller data sets. Normally, the database is divided into data sets based on usage. If a fragment contains the data that is more relevant to one location, then data is kept at that location.

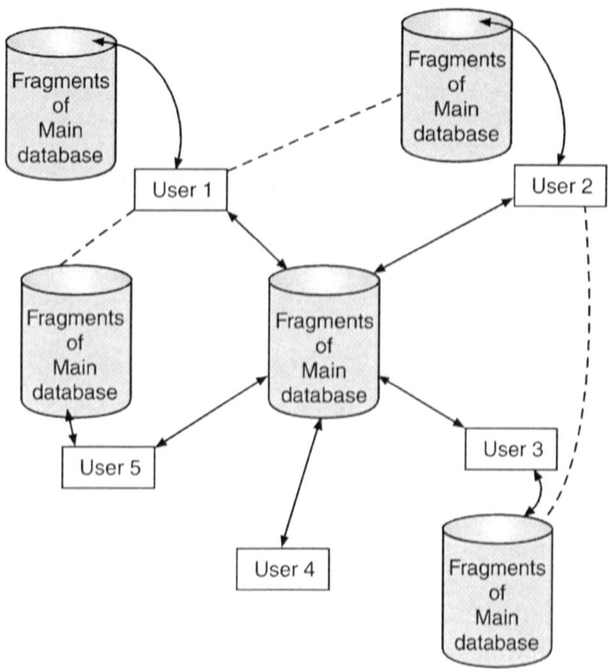

Fig. 10.21. Distributed database

Distributed database system is a collection of data with different parts under the control of separate DBMSs running on different computers. These computers are connected together and communicate with one another through various communication techniques. These system do not share the common memory, but communicates among themselves.

DDBMS can handle both local and global transactions.

The main difference between centralized and distributed database system is that, in the former, the data reside in single location, whereas in the later, the data reside in sensal locations. This distribution of data is the cause of many difficulties in transaction processing and query processing.

Distributed database are of two types :

1. Homogeneous Distributed Database
2. Heterogeneous Distributed Database

10.11.1. Homogeneous Distributed Database

In a Homogeneous distributed database, all sites have identical database management system software, are aware of one another, and agree to cooperate in processing user's requests. In such a system local sites surrender a portion of their autonomy in terms of their right to change schemas or database management system software. That software must also co-operate with other sites in exchanging information about transactions, to make transaction. Processing possible across multiple sites.

10.11.2. Heterogeneous Distributed Database

In a heterogeneous distributed database, different sites may use different schemas, and different database management system software. The sites may not be aware of one another, and they may provide only limited facilities for cooperation in transaction processing.

There are two approaches to strong the relation in the distributed database :

1. Data replication
2. Data fragmentation (Explain in Article 6.9).

Data Replication

If relation '*r*' is replicated, a copy of relation *r*, is stored in two or more sites. In the most extreme cases, we have full replication, in which a copy is stored in every site in the system.

There are a number of advantages and disadvantages to replication.

- **Availability :** If one of the sites containing relation '*r*' fails, then the relation '*r*' can be found in another site. Thus, the system can continue to process queries involving r, despite the failure of one site.
- **Increased parallelism :** In the case, where the majority of accesses to the relation 'r' result in only the reading of the relation, then several sites can process queries involving '*r*' in parallel. The more replicas of '*r*' there are the greater the chance that the needed data will be found in the site where the transaction is executing. Hence, data replication minimizes movement of data between sites.
- **Increased overhead on update :** The system must ensure that all replicas of a relation '*r*' are consistent; otherwise, erroneous computations may result. Thus, whenever *r* is updated, the update must be propagated to all sites containing replicas. The result is increased overhead.

10.11.3. Transaction Processing in DDBMS

Access to the various data items in a distributed database system is usually accomplished through transactions which must preserve the ACID properties. There are two types of transaction that we need to consider. The local transactions are those that access and update data in only one local database. The global transactions are those that access and update data in several local databases.

System Structure

Each site has its own local transaction manager, whose function is to ensure that ACID properties of those transactions that execute at that site. The various transactions managers cooperate to execute global transactions.

- **Transaction manager :** The transaction manager manages the execution of those transactions that access data stored in a local site. Note that each such transaction may be either a local transaction or part of a global transaction.
- **Transaction coordinator :** The transaction coordinator coordinates the execution of the various transaction (both local and global) inviated at that site.

Transaction in a DDBMS, access records by issuing read and write operations. Each distributed database Management system consists of a Transaction Manager (TM), a Data Manager (DM) and a scheduler.

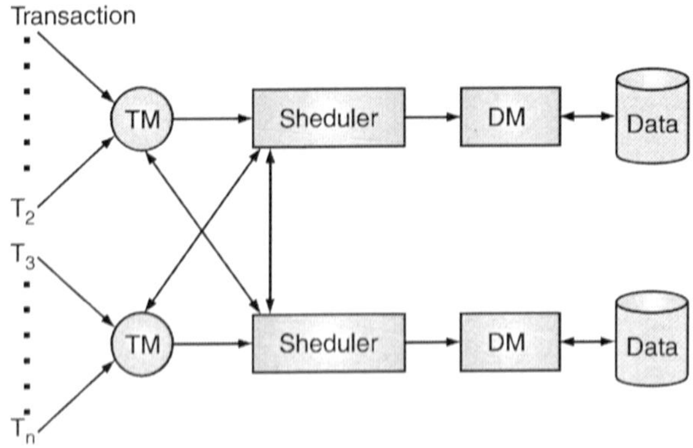

Fig. 10.22.

Transactions communicate with TMs, TMs talks to schedulers regarding each read and write of the transaction, schedulers communicate among themselves and also with DMs; and DMs manage the data.

10.11.4. The Major Advantages of DDBMS

1. DDBMS allows each site to store and maintain its own database, causing immediate and efficient access to data.
2. It allows to access the data stored at remote sites. At the same time users can retain the control to its own site to access the local data.
3. If one site is not working due to any reason (e.g., communication link goes down) the system will not be down because other sites of the network can possibly continue functioning.
4. New sites can be added to the system anytime with no or little efforts.

5. If a user needs to access the data from multiple sites then the desired query can be subdivided into sub-queries and the parts evaluated in parallel.

Disadvantages of Distributed Databases

1. Complex software is required for a distributed database environment.
2. The various sites must exchange messages and perform additional calculations to ensure proper coordination among the sites.
3. A by-product of the increased complexity and need for coordination is the additional exposure to improper updating and other problems of data integrity.
4. If the data are not distributed properly according to their usage, or if queries are not formulated correctly, response to requests for data can be extremely slow.

Summary

- Collections of operations that form a single logical unit of work are called transaction.
- Transactions have four basic properties, which are called ACID properties.
- Transaction processing system are the systems, which includes large size database and allows many users to use same database concurrently.
- The phenomenon in which a single transaction failure leads to a series of transaction rollback is known as "Cascading Rollback".
- Checkpoints is a point of synchronization between the database and the transaction log file.
- A distributed database system consists of a collection of sites connected together by the use of the some kind of communication network.

  Distributed database system can also be define as a collection of data with different parts under the control of separate DBMS's running on different computers. It is of two types : Homogeneous Distributed Database and Heterogeneous Distributed Database.
- Data fragmentation is a method in which different relations of a relational database system can be subdivided and distributed among different network sites. The union of these fragments reconstructs, the original relation S. It is of three types : Horizontal fragmentation, Vertical fragmentation, and mixed fragmentation.

Selected Exercises

1. What do you mean by transaction? Explain the transaction states.
2. What are the ACID properties? Explain in brief.
3. Explain the transaction processing system in DDBMS with figure.
4. What do you mean by data fragmentation ? Also explain its types.
5. What are various techniques of Recovery from Transaction failures ?
6. Explain log based Recovery and Checkpoints.

❑❑❑

Chapter

11 Database Security

Inside this Chapter

11.1. ELEMENTARY CONCEPTS TO DATABASE SECURITY

Confidently, integrity, and availability are the concentrations of database security. Authorization is the permission given to a user, program, or process to access an object or set of objects. The type of data access granted to a user can be read-only or read/write. Privileges specify the type of Data Manipulation Language (DML) operations that the user can perform on data.

Database integrity ensures that data in the database is correct and consistent. Database integrity mechanisms can be divided into those that support system integrity and those that enforce relational database integrity properties such as entity integrity, referential integrity, transaction integrity, and business rules. Traditional system integrity involves ensuring that data retrieved from the system is the same as the data when it was originally inserted. In addition, data must not be altered or deleted by a user who is not authorized to do so. A database must ensure that data adheres to certain business rules, as determined by the database administrator or application developer. For example, assume that a business rule says that no employee in the emp table can receive a raise greater than 20 percent of the value in the salary column. If an insert or update statement attempts to violate this integrity rule, then the statement must fail. Integrity constraints and database triggers can be used to enforce data integrity rules within a database.

Referential integrity rules form the basis of data integrity in relational databases. Referential integrity ensures that data entered in one table, which references another table in a relational database, and the rules that govern this data are in conformance with the referenced values. Referential integrity includes rules that dictate the types of data manipulation allowed on referenced values, and these rules are used to determine the effect of data manipulation on dependent values.

Database Security

Threats and risks to databases have increased and therefore, the need for securing databases has also increased. Let's learn about the basic facets of database security, including assurance, integrity, availability, and confidentiality.

The majority of the companies store sensitive data in databases. However, database security is sometimes not given as much thought and effort as other areas of computer security. Hackers have been able to target large databases in recent years to obtain sensitive information like credit card numbers and other personal information. It is important to protect databases against these risks, and this is where database security comes into place.

Database security can_be defined as a system or process by which the "Confidentiality, Integrity, and Availability," or CIA, of the database can be protected. Unauthorized entry or access to a database server signifies a loss of confidentiality; unauthorized alteration to the available data signifies loss of integrity; and lack of access to database services signifies loss of availability. Loss of one or more of these basic facets will have a significant impact on the security of the database.

For an illustration of this concept, imagine that the website of a company contains information like who they are, what they do, and what prospective customers have to do to contact them for their queries. In this case, the availability of the database services is more important when compared with other factors like the confidentiality or integrity of the database security.

For a company that sells products or goods online, however, confidentiality and integrity are more important as customers use their credit cards to buy goods online only when the site is available.

Another factor needs to be addressed when examining database security and that is "Assurance."

What is database assurance? Take for example, a web application that acts as a frontend to a database server. If the web application that is selling online goods is vulnerable to cross-site-scripting, the chances of people not trusting the website becomes greater. When customers lose trust or assurance in the company, this may consequently lead to loss in the business.

Databases are susceptible to other vulnerabilities like poor password management, SQL injection, leakage of data, and improper error handling apart from cross site scripting. Hackers try to attack databases that are configured poorly. Hackers take advantage of these database weaknesses to exploit the database vulnerabilities.

The risks involved with the database are not the same for every database present in the organization; therefore, security controls or measures to these databases differ. As there are various databases like Oracle, SQL and Access, different types of database security solutions are also available in the market. One needs to assess the risk for the database involved and mitigate the risk by designing and implementing appropriate database security solutions. If security is the key driver for database configuration, the data will remain safe and secure.

Database security entails allowing or disallowing user actions on the database and the objects within it. Oracle uses schemas and security domains to control access to data and to restrict the use of various database resources.

Oracle provides comprehensive discretionary access control. Discretionary access control regulates all user access to named objects through privileges. A privilege is permission to access a named object in a prescribed manner; for example, permission to query a table. Privileges are granted to users at the discretion of other users.

Database Users and Schemas

Each Oracle database has a list of user names. To access a database, a user must use a database application and attempt a connection with a valid user name of the database. Each user name has an associated password to prevent unauthorized use.

Security Domain

Each user has a security domain—a set of properties that determine such things as:

- The actions (privileges and roles) available to the user
- The tablespace quotas (available disk space) for the user
- The system resource limits (for example, CPU processing time) for the user

Each property that contributes to a user's security domain is discussed in the following sect

Database security requirements arise from the need to protect data from accidental loss and corruption as well as from deliberate unauthorized attempts to access or alter that data. Security concerns also include protecting against undue delays in accessing or using data, interference, and in extreme cases, denial of service.

11.2. ORACLE ADVANCED SECURITY ARCHITECTURE

Oracle Advanced Security provides a comprehensive suite of security features to protect enterprise networks and securely extend corporate networks to the Internet. It provides a single source of integration with network encryption and authentication solutions, single sign-on services, and security protocols. By integrating industry standards, it delivers unparalleled security to an Oracle networking environment. Oracle Advanced Security complements an Oracle server or client installation with advanced security features.

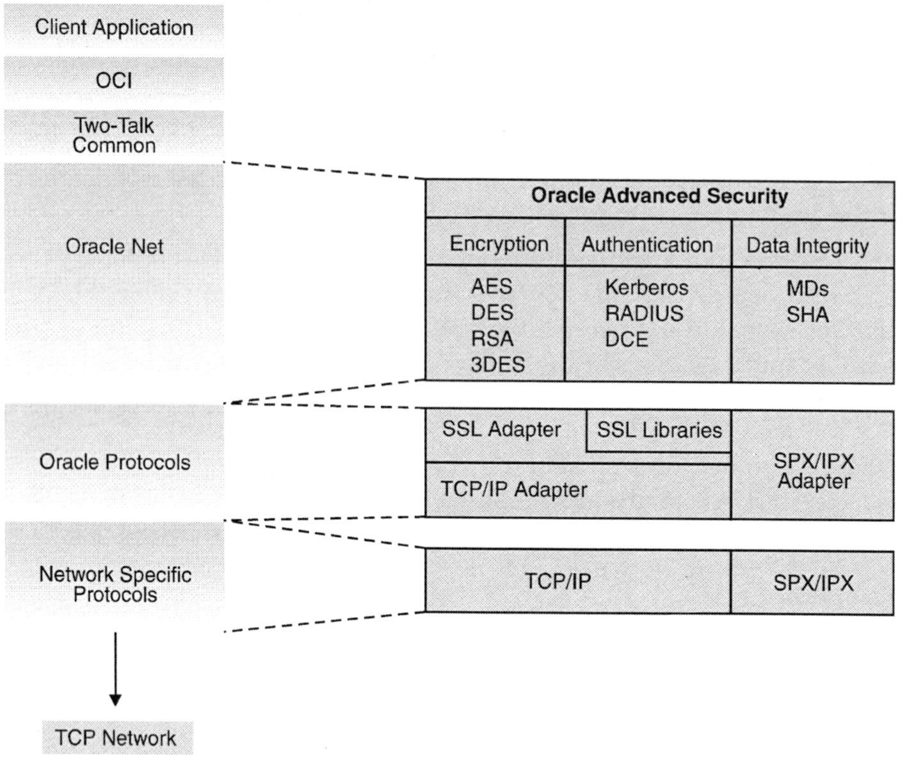

Fig. 11.1. Oracle Advanced Security in an Oracle Networking Environment

Oracle Advanced Security supports authentication through adapters that are similar to the existing Oracle protocol adapters. As shown in Figure 11.2, authentication adapters integrate under the Oracle Net interface and allow existing applications to take advantage of new authentication systems transparently, without any changes to the application.

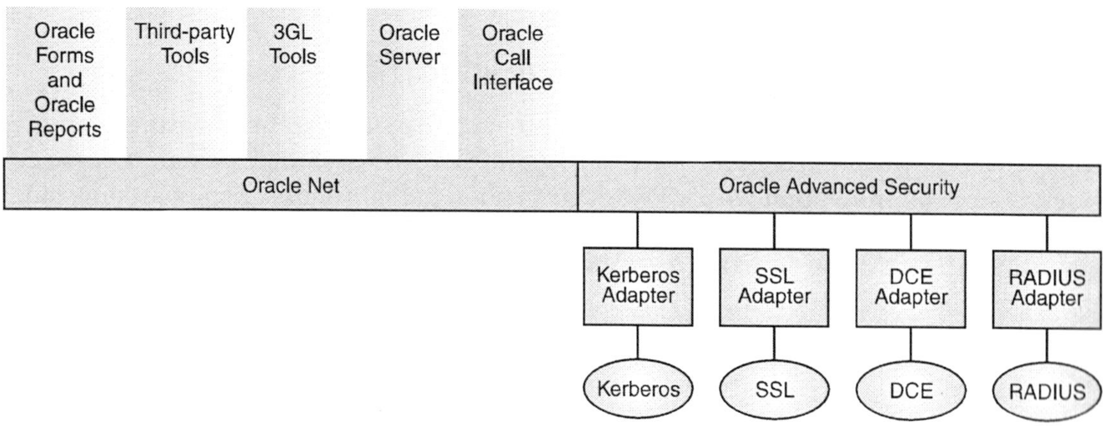

Fig. 11.2. Oracle Net with Authentication Adapters

11.3. SOLVING SECURITY CHALLENGES WITH ORACLE ADVANCED SECURITY

To solve enterprise computing security problems, Oracle Advanced Security provides various options to ensure industry standards-based data privacy, integrity, authentication, single sign-on, and access authorization. For example, you can configure either Oracle Net native encryption or Secure Sockets Layer (SSL) for data privacy. Oracle Advanced Security also provides the choice of several strong authentication methods, including Kerberos, smart cards, and digital certificates.

Oracle Advanced Security provides the following security features:

- Data Encryption
- Strong Authentication

11.4. DATA ENCRYPTION

Sensitive information that travels over enterprise networks and the Internet can be protected by encryption algorithms. An encryption algorithm transforms information into a form that can be deciphered only with a decryption key. Figure 11.3 shows how encryption works to ensure the security of a transaction. For example, if a manager approves a bonus, this data should be encrypted when sent over the network to avoid eavesdropping. If all communication between a client, database, and an application server is encrypted, then when the manager sends the bonus amount to the database, the information is protected.

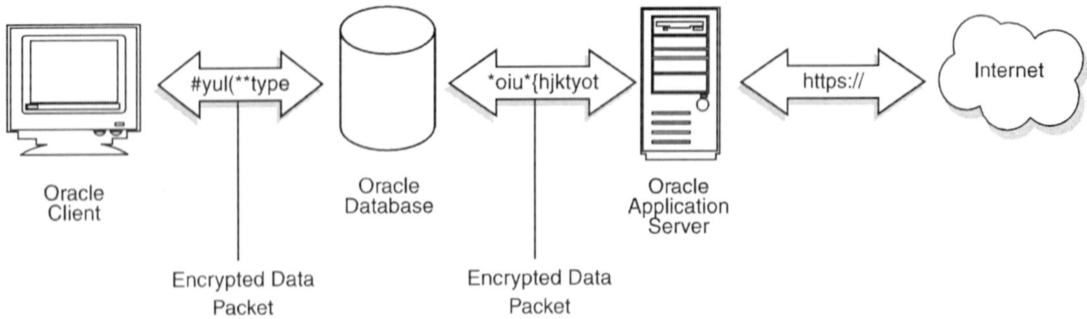

Fig. 11.3. Encryption

This section discusses the following topics:

- Supported Encryption Algorithms
- Data Integrity
- FIPS

Supported Encryption Algorithms

Oracle Advanced Security provides the following encryption algorithms to protect the privacy of network data transmissions:

- RC4 Encryption
- DES Encryption
- Triple-DES ODES) Encryption
- Advanced Encryption Standard

Selecting the network encryption algorithm is a user configuration option that provides varying levels of security and performance for different types of data transfer. Prior versions of Oracle Advanced Security provided three editions: Domestic, Upgrade, and Export—each with different key lengths, Oracle Advanced Security 10 g Release 1 (10.1) contains a complete complement of the available encryption algorithms and key lengths, previously only available in the Domestic edition. Users deploying prior versions of the product can obtain the Domestic edition for a specific product release.

11.4.1. RC4 Encryption

The RC4 encryption module uses the RC4 encryption algorithm created by RSA Security, Inc. Using a secret, randomly-generated key that is unique to each session, all network traffic is fully safeguarded, including all data values, SQL statements, and stored procedure calls and results. The client, server, or both, can require or request the use of the encryption module to guarantee data protection. The optimized implementation supported by Oracle provides a high degree of security for a minimal performance penalty. For the RC4 algorithm, Oracle provides encryption key lengths of 40-bits, 56-bits, 128-bits, and 256-bits.

11.4.2. DES Encryption

Oracle Advanced Security implements the U.S. Data Encryption Standard (DES) algorithm with a standard, optimized, 56-bit key encryption, and also provides DES40, a 40-bit version, for backward compatibility.

11.4.3. Triple-DES (3DES) Encryption

Oracle Advanced Security also supports Triple-DES encryption (3DES), which encrypts message data with three passes of the DES algorithm. 3DES provides a high degree of message security, but with considerable performance penalty. The magnitude of penalty depends on the speed of the processor performing the encryption. 3DES typically takes three times as long to encrypt a data block as compared to the standard DES algorithm. 3DES is available in two-key and three-key versions with effective key lengths of 112 bits and 168 bits, respectively. Both versions operate in outer Cipher Block Chaining (CBC) mode.

11.4.4. Advanced Encryption Standard

Approved by the National Institute of Standards and Technology (NIST) in Federal Information Processing Standards (FIPS) Publication 197, Advanced Encryption Standard (AES) is a new cryptographic algorithm standard developed to replace DES. AES is a symmetric block cipher that can process data blocks of 128 bits, using cipher keys with lengths of 128, 192, and 256 bits, which are referred to as AES-128, AES-192, and AES-256, respectively. All three versions operate in outer CBC mode.

Data Integrity

To ensure the integrity of data packets during transmission, Oracle Advanced Security can generate a cryptographically secure message digest, using MD5 or SHA-1 hashing algorithms, and include it with each message sent across a network.

Data integrity algorithms add little overhead, and protect against the following attacks:

- Data modification
- Deleted packets
- Replay attacks

FIPS

Oracle Advanced Security Release 8.1.6 has been validated under U.S. FIPS 140-1 at the Level 2 security level. This provides independent confirmation that Oracle Advanced Security conforms to federal government standards.

11.5. STRONG AUTHENTICATION

Authentication is used to prove the identity of a user. Authentication, without which there can be little confidence in network security, is imperative in distributed environments. Passwords are the most common means of authentication. Oracle Advanced Security provides strong authentication with Oracle authentication adapters that support various third-party authentication services, including SSL with digital certificates.

Figure 11.4. shows user authentication with an Oracle database configured to use a third-party authentication server. Having a central facility to authenticate all the members of a network (clients to servers, servers to servers, and users to both clients and servers) is one effective way to address the threat of network nodes falsifying their identities.

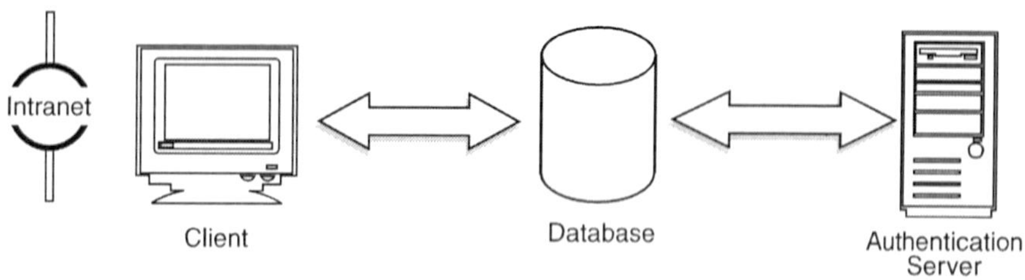

Fig. 11.4. Strong Authentication with Oracle Authentication Adapters

11.6. DATABASE SECURITY POLICIES

This section briefly introduces security policies. It covers:

- Security Threats and Countermeasures
- What Information can Security Policies Cover

Security Threats and Countermeasures

An organization should create a well-documented security policy. This policy should enumerate the security threats the organization is trying to guard against and the specific measures the organization must take when it encounters these threats. Security threats can be addressed by the following measures:

- Procedural, such as requiring data center employees to display security badges

- Physical, such as securing computers containing critical information in restricted-access facilities
- Technical, such as implementing strong authentication requirements for critical business systems
- Personnel-related, such as performing background checks or vetting key personnel

The organization should consider whether the response to a threat is procedural, physical, technical, personnel-related, or a combination of these. For example, one possible security threat is the disruption of critical business systems caused by a malicious person damaging a computer. A physical response to this threat is to secure key business computers in a restricted-access facility. A procedural response is to create system backups at regular intervals. Personnel measures could include background checks on employees who access or manage key business systems.

What Information can Security Policies Cover

In addition to addressing requirements unique to the organization environment, the organization should also design and implement technical measures in its information security policies to address generic issues, as listed in Table 4-1:

Table 4.1 Issues and Actions Governing Security Policies

Security Concern/Practice	Recommended Actions
Establish and maintain application-level security	Attach privileges and roles to each application, Ensure that users cannot misuse these roles and privileges when they are not using the application. Base the use of roles on user-specific criteria, such as a user connecting only from a particular IP address, or only through a particular application tier.
Manage privileges and attributes (system/object/user)	Permit only a limited set of users to access, process, or alter data or to run a particular type of SQL statement or to access another user's object. Apply the required limitations on user access to objects or actions on objects, such as schemas, tables, or rows, or resources, such as system usage time (CPU, connect, or idle times).
Create, manage, and control roles (database, enterprise)	Create named groups of privileges called roles to facilitate granting them to users, including previously named groups.
Establish the granularity of access control desired	Set up session-based attributes in a secure manner. For example, store user attributes (user name, employee number, and so on) to be retrieved later in the session, enabling fine-grained access control.

Security Concern/Practice	Recommended Actions
	Create security policy functions and attach them to critical or sensitive tables, views, or synonyms used by an application. DML statements on such objects are then modified dynamically, and transparently to the user, to disable inappropriate access.
Establish and manage the use of encryption	Use Secure Socket Layer (SSL) connections, well-established encryption suites, or PKI certificates for critical/sensitive transmissions/applications.
Establish and maintain security in 3-tier applications	Preserve user identity through a application tier to the database. Avoid the overhead of separate database connections by proxying user identities (and credentials such as a password or certificate) through the application tier to the database.
Control query access, data misuse, and intrusions	Monitor query access based on specific content or row to detect data misuse or intrusions. Use proxy authentication to support auditing of proxied user connections. Use Regular Auditing and Fine-Grained Auditing to detect unauthorized or inappropriate access or actions.

Authentication by the Oracle Database

Oracle can authenticate users attempting to connect to a database, by using information stored in that database.

To set up Oracle to use database authentication, you create each user with an associated password that must be supplied when the user attempts to establish a connection. This process prevents unauthorized use of the database because the connection will be denied if the user provides an incorrect password. Oracle stores a user,s password in the data dictionary in an encrypted format to prevent unauthorized alteration, but a user can change his own password at any time.

To establish which authentication protocols are allowed by the client or database, a DBA can explicitly set the SQLNET . ALLOWED_LOGON_VERsion parameter in the sqlnet. ora server file. Then each connection attempt is tested, and if the client or server does not meet the minimum version specified by its partner, authentication fails with an ORA-28040 error. The parameter can take the values 10, 9, or 8 (default), which represent the database server versions. Oracle recommends the value 10.

Database authentication includes the following facilities:

- Password Encryption While Connecting
- Account Locking
- Password Lifetime and Expiration

- Password History
- Password Complexity Verification »

Password Encryption While Connecting

Passwords are always automatically and transparently encrypted during network (client/ server and server/server) connections. This is done by using a modified DBS (Data Encryption Standard) or 3DES algorithm before sending them across the network.

Account Locking

Oracle can lock a user's account after a specified number of consecutive failed log-in attempts. You can configure the account to unlock automatically after a specified time interval or by database administrator intervention.

Use the CREATE PROFILE statement to specify the number of failed login attempts allowed before the account locks and the duration for which the account remains locked before it unlocks automatically.

The DBA can also lock accounts manually. In these cases, the accounts cannot unlock automatically and must be unlocked explicitly by the DBA.

Password Lifetime and Expiration

The database administrator can specify a lifetime for passwords, after which they expire and must be changed before account login is permitted. A grace period can be specified, during which each attempt to log in to the database account receives a warning message to change the password. If it is not changed by the end of that period, the account is locked. No further login to that account is allowed without assistance by the DBA.

The DBA can also manually set the password state to expired, causing the user's account status to change to expired. The user or the DBA must then change the password before the user can log in to the database.

Password History

The password history option checks each newly specified password to ensure that a password is not reused for a specified amount of time or for a specified number of password changes. The database administrator can configure the rules for password reuse with CREATE PROFILE statements.

Password Complexity Verfication

Password complexity verfication checks that each password is complex enough to provide reasonable protection against intruders who try to break into the system by guessing passwords.

The sample Oracle password complexity verification routine (PL/SQL script UTLPWDMG.SQL, which sets the default profile parameters) specifies that a password must:

- Be a minimum of four characters in length.
- Xot be the same as the user ID.
- Include at least one alphabet character, one numeric character, and one punctuation mark.

- Not match any word on an internal list of simple words, such as welcome, account, database, and user.
- Differ from the old password by at least three characters.

11.7. BACKUP AND RECOVERY: BASIC CONCEPTS

The physical structures of the database and the role each plays in the database recovery process are what determine the forms of backup and recovery available through user-managed techniques and through RMAN.

Physical Database Structures Used in Recovering Data

The files and other structures that make up an Oracle database store data and safeguard it against possible failures. This section introduces each of the physical structures that make up an Oracle database and their role in the reconstruction of a database from backup. This section contains these topics:

- Datafiles and Data Blocks
- Redo Logs
- Undo Segments
- Control Files

11.7.1. Datafiles and Data Blocks

An Oracle database consists of one or more logical storage units called tablespaces. Each tablespace in an Oracle database consists of one or more files called datafiles, physical files under the host operating system in which the database is running.

A database's data is collectively stored in the datafiles that constitute each tablespace of the database. The simplest Oracle database would have one tablespace, stored in one datafile. The datbase manages the storage space in the datafiles of a database in units called data blocks. A data block is the smallest unit of data used by a database. Data blocks are the smallest units of storage that the database can use or allocate.

Modified or new data is not written to datafiles immediately. Updates are buffered in memory and written to datafiles at intervals. If a database has not gone through a normal shutdown (that is, if it is open, or exited abnormally, as in an instance failure or a SHUTDOWN ABORT) then there are typically changes in memory that have not been written to the datafiles. Datafiles that were restored from backup, or were not closed during a **consistent shutdown**, are typically not completely up to date.

Copies of the datafiles of a database are a critical part of any backup.

11.7.2. Redo Logs

Redo logs record all changes made to a database's data files. With a complete set of redo logs and an older copy of a datafile, the database can reapply the changes recorded in the redo logs to re-create the database at any point between the backup time and the end of the last redo log. Each time data is changed in the database, that change is recorded in the **online redo log** first, before it is applied to the datafiles. An Oracle database requires at least two **online redo log groups**, and in each group there is at least one **online redo log member**, an individual redo log file where the changes are recorded.

At intervals, the database rotates through the online redo log groups, storing changes in the **current online redo log** while the groups not in use can be copied to an archive location, where they are called **archived redo logs** (or, collectively, the **archived redo log**). You can run your database in ARCHIVELOG mode (in which this archiving of redo log files is enabled) or NOARCHIVELOG mode (in which redo log files are simply overwritten).

Preserving the archived redo log is a major part of most backup strategies, as they contain a record of all updates to datafiles. Backup strategies often involve copying the archived redo logs to disk or tape for longer-term storage. Running in NOARCHIVELOG mode limits your data recovery options.

11.7.3. Control Files

The control file contains a crucial record of the physical structures of the database and their status. Several types of information stored in the control file are related to backup and recovery:

- Database information (RESETLOGS SCN and time stamp)
- Tablespace and datafile records (filenames, datafile checkpoints, read/write status, offline ranges)
- Information about redo threads (current online redo log)
- Log records (log sequence numbers, SCN range in each log)
- A record of past RMAN backups
- Information about corrupt datafile blocks

The recovery process for datafiles is in part guided by status information in the control file, such as the database checkpoints, current online redo log file, and the datafile header checkpoints for the datafiles. Loss of the control file makes recovery from a data loss much more difficult.

11.7.4. Undo Segments

In general, when data in a datafile is updated, "before images" of that data are written into **undo segments**. If a transaction is rolled back, this undo information can be used to restore the original datafile contents.

In the context of recovery, the undo information is used to undo the effects of uncommitted transactions, once all the datafile changes from the redo logs have been applied to the datafiles. The database is actually opened before the undo is applied.

You should not have to concern yourself with undo segments or manage them directly as part of your backup and recovery process.

Creating a Backup and Recovery Plan

Data backup is an insurance plan. Important files are accidentally deleted all the time. Mission-critical data can become corrupt. Natural disasters can leave your office in ruin. With a solid backup and recovery plan, you can recover from any of these. Without one, you're left with nothing to fall back on.

Figuring Out a Backup Plan

It takes time to create and implement a backup and recovery plan. You'll need to figure out what data needs to be backed up, how often the data should be backed up, and more. To help you create a plan, consider the following:

11.8. THE BASIC TYPES OF BACKUP

There are man techniques for backing up files. The techniques you use will depend on the type of data you're backing up, how convenient you want the recovery process to be, and more.

If you view the properties of a file or directory in Windows Explorer, you'll note an attribute called Archive. This attribute often is used to determine whether a file or directory should be backed up. If the attribute is on, the file or directory may need to be backed up. The basic types of backups you can perform include

- **Normal/full backups** All files that have been selected are backed up, regardless of the setting of the archive attribute. When a file is backed up, the archive attribute is cleared. If the file is later modified, this attribute is set, which indicates that the file needs to be backed up.
- **Copy backups** All files that have been selected are backed up, regardless of the setting of the archive attribute. Unlike a normal backup, the archive attribute on files isn't modified. This allows you to perform other types of backups on the files at a later date.
- **Differential backups** Designed to create backup copies of files that have changed since the last normal backup. The presence of the archive attribute indicates that the file has been modified and only files with this attribute are backed up. However, the archive attribute on files isn't modified. This allows you to perform other types of backups on the files at a later date.
- **Incremental backups** Designed to create backups of files that have changed since the most recent normal or incremental backup. The presence of the archive attribute indicates that the file has been modified and only files with this attribute are backed up. When a file is backed up, the archive attribute is cleared. If the file is later modified, this attribute is set. which indicates that the file needs to be backed up.
- **Daily backups** Designed to back up files using the modification date on the file itself. If a file has been modified on the same day as the backup, the file will be backed up. This technique doesn't change the archive attributes of files.

In your backup plan you'll probably want to perform full backups on a weekly basis and supplement this with daily, differential, or incremental backups. You may also want to create an extended backup set for monthly and quarterly backups that includes additional files that aren't being backed up regularly.

Tip You'll often find that weeks or months can go by before anyone notices that a file or data source is missing. This doesn't mean the file isn't important. Although some types of data aren't used often, they're still needed. So don't forget that you may also want to create extra sets of backups for monthly or quarterly periods, or both, to ensure that you can recover historical data over time.

Differential and Incremental Backups

The difference between differential and incremental backups is extremely important. To understand the distinction between them, examine Table 14-1. As it shows, with differential backups you back up all the files that have changed since the last full backup (which means that the size of the differential backup grows over time). With incremental backups, you only back up files that have changed since the most recent full or incremental backup (which means the size of the incremental backup is usually much smaller than a full backup).

11.9. AUTHENTICATION VS. AUTHORIZATION

It is easy to confuse the mechanism of authentication with that of authorization. In many host-based systems (and even some client/server systems), the two mechanisms are performed by the same physical hardware and, in some cases, the same software.

It is important to draw the distinction between these two mechanisms, however, since they **can** (and, one might argue, **should**) be performed by separate systems.

What, then, distinguishes these two mechanisms from one another?

Authentication is the mechanism whereby systems may securely identify their users. Authentication systems provide an answers to the questions:

- Who is the user?
- Is the user really who he/she represents himself to be?

An authentication system may be as simple (and insecure) as a plain-text password challenging system (as found in some older PC-based FTP servers) or as complicated as the *Kerberos* system described elsewhere in these documents. In all cases, however, authentication systems depend on some unique bit of information known (or available) only to the individual being authenticated and the authentication system—a *shared secret*. Such information may be a classical password, some physical property of the individual (fingerprint, retinal vascularization pattern, etc.), or some derived data (as in the case of so-called *smartcard* systems). In order to verify the identity of a user, the authenticating system typically challenges the user to provide his unique information (his password, fingerprint, etc.)—if the authenticating system can verify that the shared secret was presented correctly, the user is considered authenticated.

Authorization, by contrast, is the mechanism by which a system determines what level of access a particular authenticated user should have to secured resources controlled by the system. For example, a database management system might be designed so as to provide certain specified individuals with the ability to retrieve information from a database but not the ability to change data stored in the datbase. while giving other individuals the ability to change data. Authorization systems provide answers to the questions:

Selected Exercises

1. Explain the concepts of database Security ?
2. How can the security be enhanced ?
3. What are the various security challenges with oracle Advance Database ?
4. Explain the different types of Encryption ?
5. What are the various Database Security policies Explain ?
6. Explain the various important Backup Techniques ?
7. What are the different types of Backup ? Explain.
8. Mention the important difference between Authentication and Authorization.

❑❑❑

A PRACTICAL APPROACH TO SQL

A PRACTICAL APPROACH TO SQL

Objective 1 : Create tables and specify the Questionnaires in SQL

Theory and Concepts :

Introduction about SQL

SQL (Structured Questionary Language) is a nonprocedural language, you specify what you want, not how to get it. A block structured format of English key words is used in this Questionary language. It has the following components.

DDL (Data Definition Language)

The SQL DDL provides command for defining relation schemes, deleting relations and modifying relation schema.

DML (DATA Manipulation Language)

It includes commands to insert tuples into, delete tuples from and modify tuples in the database.

View Definition

The SQL DDL includes commands for defining views.

Transaction Control-SQL includes for specifying the beginning and ending of transactions.

Embedded SQL and Dynamic SQL

Embedded and Dynamic SQL definite how SQL statements can be embedded with in general purpose programming languages, such as C, C++, JAVA, COBOL, Pascal and Fortran.

Integrity

The SQL DDL include commands for specifying integrity constrains that the data stored in the database must specify. Updates that violent integrity constraints are allowed.

Authorization

The SQL DDL includes commands for specifying access rights to relations and views.

Data Definition Language

The SQL DDL allows specification of not only a set of relations but also information about each relation, including.

- Schema for each relation
- The domain of values associated with each attribute.

- The integrity constraints.
- The set of indices to be maintained for each relation.
- The security and authorization information for each relation.
- The physical storage structure of each relation on disk.

Domain types in SQL

The SQL standard supports a variety of built in domain types, including

- Char (*n*)–A fixed length character length string with user specified length.
- Varchar (n)–A variable character length string with user specified maximum length *n*.
- Int–An integer.
- Small integer–A small integer.
- Numeric (*p*, *d*)–A fixed point number with user defined precision.
- Real, double precision–Floating point and double precision floating point numbers with machine dependent precision.
- Float (*n*)–A floating point number, with precision of at least *n* digits.
- Date–A calender data containing a (four digit) year, month and day of the month.
- Time–The time of day, in hours, minutes and second, *e.g.*, Time '09:30:00'.
- Number–Number is used to store numbers (fixed or floating point).

DDL Statement for Creating a Table

Syntax

Create table tablename

(colourname datatype(size), columnname datatype(size));

Creating a table from a table-

Syntax

CREATE TABLE TABLENAME

[(coloumnname, columnname, ...)]

AS SELECT columnname, columnname, ... FROM tablename;

Insertation of data into tables

Syntax

INSERT INTO tablename

[(columnname, columnname, ...)]

Values (expression, expression);

Inserting Data into a Table from Another Table :

Syntax

INSERT INTO tablename

SELECT columnname, columnname, ...

FROM tablename;

Insertion of Selection Data into a Table from another Table

Syntax

```
INSERT INTO tablename
SELECT columnname, columnname ......
FROM tablename
WHERE columnname=expression;
```

Retrieving Data from the Tables

Syntax

```
SELECT * FROM tablename;
```

The Retrieving of Specific Columns from a Table

Syntax

```
SELECT columnname, columnname, .....
FROM tablename;
```

Elimination of Duplicates from the Select Statement

Syntax

```
SELECT DISTINCT columnname, columnname
FROM tablename;
```

Selecting a Data set from Table Data

Syntax

```
SELECT columnname, columnname
FROM tablename
WHERE searchcondition;
```

Q. 1 Create the following tables :

(*i*) **Client_master**

columnname	datatype	size
client_no	varchar2	6
name	varchar2	20
address1	varchar2	30
address2	varchar2	30
city	varchar2	15
state	varchar2	15
pincode	number	6
bal_due	number	10,2

(ii) Product_master

columnname	datatype	size
Product_no	varchar2	
Description	varchar2	
Profit_percent	number	
Unit_percent	varchar2	
Qty_on_hand	number	
Reorder_|v|	number	
Sell_price	number	
Cost_price	number	

Q.2 Insert the following data into their respective tables :

Client no	Name	City	Pin code	State	Bal.due
0001	Ivan	Bombay	400054	Maharashtra	15000
0002	Vandana	Madras	780001	Tamilnadu	0
0003	Pramada	Bombay	400057	Maharashtra	5000
0004	Basu	Bombay	400056	Maharashtra	0
0005	Ravi	Delhi	100001		2000
0006	Rukmani	Bombay	400050	Maharashtra	0

Data for Product Master :

Product No.	Description	Profit %	Unit Percent	Qty measured	Reorder on hand|v|	Sell price	Cost price
P00001	1.44 floppies	5	piece	100	20	525	500
P03453	Monitors	6	piece	10	3	12000	11200
P06734	Mouse	5	piece	20	5	1050	500
P07865	1.22 floppies	5	piece	100	20	525	500
P07868	Keyboards	2	piece	10	3	3150	3050
P07885	CD Drive	2.5	piece	10	3	5250	5100
P07965	540 HDD	4	piece	10	3	8400	8000
P07975	1.44 Drive	5	piece	10	3	1050	1000
P08865	1.22 Drive	5	piece	2	3	1050	1000

Q.3 On the basis of above two tables answer the following questionnaires :

(*i*) Find out the names of all the clients.

(*ii*) Retrieve the list of names and cities of all the clients.

(*iii*) List the various products available form the product_master table.

(*iv*) List all the clients who are located in Bombay.

(*v*) Display the information for client no 0001 and 0002.

(*vi*) Find the products with description as '1.44 drive' and '1.22 Drive'.

(*vii*) Find all the products whose sell price is greater than 5000.

(*viii*) Find the list of all clients who stay in city 'Bombay' or city 'Delhi' or 'Madras'.

(*ix*) Find the product whose selling price is greater than 2000 and less than or equal to 5000.

(*x*) List the name, city and state of clients not in the state of 'Maharashtra'.

Objective 2 : To Manipulate the Operations on the table.

DML (Data Manipulation Language) Data manipulation is

- The retrieval of information stored in the database.
- The insertion of new information into the database.
- The detection of information from the database.
- The modification of information stored by the appropriate data model. There are basically two types.
  - (*i*) **Procedural DML :** require a user to specify what data are needed and how to get those data.
  - (*ii*) **Non Procedural DML :** require a user to specify what data are needed without specifying how to get those data.

Updating the Content of a Table :

In creation situation we may wish to change a value in table without changing all values in the tuple. For this purpose the update statement can be used.

Update table name

Set columnname=experision, columnname=expression ......

Where columnname=expression;

Detection Operation :

A delete requisitions is expressed in much the same way as Questionary. We can delete whole tuple (rows) we can delete values on only particulars attributes.

Deletion of All Rows

Syntax

Delete from tablename :

Deletion of Specified Number of Rows

Syntax

Delete from table name

Where search condition ;

Computation in expression lists used to select data

+	Addition	– Subtraction
*	Multiplication	** exponentiation
/Division		( ) Enclosed operation

Renaming columns used with Expression Lists : The default output column names can be renamed by the user if required.

Syntax :

Select column name	result_columnname,
Columnname	result_columnname,

From table name;

Logical Operators :

The logical operators that can be used in SQL sentenced are

And	all of must be included
Or	any of may be included
NOT	none of could be included

Range Searching : Between operation is used for range searching.

Pattern Searching :

The most commonly used operation on string is pattern matching using the operation 'like' we described patterns by using two special characters.

- Percent (%) ; the % character matches any substring we consider the following examples.
- 'Perry %' matches any string beginning with perry
- '% idge % matches any string containing' idge as substring.
- '- - -' matches any string exactly three characters.
- '- - - % matches any string of at least of three characters'

Oracle Functions

Functions are used to manipulate data items and return result. Function follow the format of function_name (argument 1, argument2..). An arrangement is user defined variable or constant. The structure of function is such that it accepts zero or more arguments.

Examples :

Avg return average value of n.

Syntax

Avg ([distinct/all]n)

Min return minimum value of expr.

Syntax

MIN (distinct/all) expr)

Count Returns the no of rows where expr is not null.

Syntax

Count ([distinct/all]expr)

Count (*) Returns the no rows in the table, including duplicates and those with nulls.

Max Return max value of expr.

Syntax :

Max ([distinct/all]expr)

Sum Returns sum of values of n.

Syntax :

Sum ([distinct/all]n).

Sorting of data in table

Syntax

Select columnname, columnname

From table

Order by columnname;.

Question 1. Using the table master and product master answer the following Questionnaries.

(*i*) Change the selling price of 1.44 floppy drive to Rs. 1150.00.

(*ii*) Delete the record with client 0001 from the client master table.

(*iii*) Change the city of client_no '0005' to Bombay.

(*iv*) Change the bal_due of client_no '0001, to 1000.

(*v*) Find the products whose selling price is more than 1500 and also find the new selling price as original selling price *15.

(*vi*) Find out the clients who stay in a city whose second letter is a.

(*vii*) Find out the name of all clients having 'a' as the second letter in their names.

(*viii*) List the products in sorted order of their description.

(*ix*) Count the total number of orders

(*x*) Calculate the average price of all the products.

(*xi*) Calculate the minimum price of products.

(*xii*) Determine the maximum and minimum prices. Rename the title as 'max_price' and min_price respectively.

(*xii*) Count the number of products having price greater than or equal to 1500.

Objective 3 : To implement the restrictions on the table.

Data Constraints : Besides the cell name, cell length and cell data type there are other parameters i.e. other data constraints that can be passed to the DBA at check creation time. The constraints can either be placed at column level or at the table level.

(*i*) Column Level Constraints : If the constraints are defined along with the column definition, it is called a column level constraint.

(*ii*) Table Level Constraints : If the data constraint attached to a specify cell in a table reference the contents of another cell in the table then the user will have to use table level constraints.

Null Value Concepts : While creating tables if a row locks a data value for particular column that value is said to be null. Column of any data types may contain null values unless the column was defined as not null when the table was created.

Syntax

Create Table Tablename

(Columnname data type (size) not null....)

Primary Key : Primary key is one or more columns is a table used to uniquickly identity each row in the table. Primary key values must not be null and must be uniQuestoin across the column. A multicolumn primary key is called composite primary key.

Syntax : Primary Key as a Column Constraint

Create table tablename

(columnname datatype (size), columnname datatype (size)...

Primary key (columnname, columnname));

UniQuestion Key Concept : A UniQuestion is similar to a primary key except that the purpose of a uniQuestion key is to ensure that information in the column for each record is uniQuestion as with telephone or devices license numbers. A table may have many uniQuestion keys.

Syntax : UniQuestion as a Column Constraint

Create table table name

(columnname datatype (size) uniQuestion);

UniQuestion as table constraint :

Create table tablename

(columnname datatype (size), columnname datatype (size) ... uniQuestion

(columnname, columnname));

Default Value Concept : At the line of cell creation a default value can be assigned to it. When the user is loading a record with values and leaves this cell empty, the DBA will automatically load this cell with the default value specified. The data type of the default value should match the data type of the column.

Syntax :

Create table tablename

(columnname datatype (size) default value, ...);

Foreign Key Concept : Foreign key represents relationship between tables. A foreign key is column whose values are derived from the primary key of the same of some other table, the existence of foreign key implies that the table with foreign key is related to the primary key table form which the foreign key is derived. A foreign key must have corresponding primary key value in the primary key table to have meaning. Foreign key as a column constraint.

Syntax

Create table name

(columnname datatype (size) reference another table name);

Foreign Key as a Table Constraint :

Syntax :

Create table tablename

(columnname datatype (size) CONSTRAINT constraintname)

Check (expression));

Question 2. Create the following tables :

(i) Sales_master

Columnname	Datatype	Size	Attributes
Sales_no	varchar 2	6	Primary key/first letter must start with 's'
Sal_name	varchar2	20	Not null
Address	varchar2		Not null
City	varchar2	20	
State	varchar2	20	
Pincode	Number	6	
Sat_amt	Number	8.2	Not null, cannot be 0
Tgt_to_get	Number	6.2	Not null, cannot be 0
Vtd_sales	Number	6.2	Not null, cannot be 0
Remarks			

(ii) Sales_order

Columnname	Datatype	Size	Attributes
S_order_no	varchar2	6	Primary/first letter must be 0
S_order_date	Date	6	Primary key reference client no of client_master table.
Client_no	Varchar2	25	
Dely_add	Varchar2	6	
Salesman_no	Varchar2	6	Foreign key reference Salesman_no of salesman_master table
Dety_type	Char	1	Delivery part (p)/full (f), default f
Billed_yn	Char	1	
Dely_date	Date		Can not be less than s_order_date
Order_status	Varchar2	10	Values ('in process'; 'fulfilled'; 'back order'; canceled

(i) Sales_order_details

Column	Datatype	Size	Attributes
S_order_no	Varchar2	6	Primary key/foreign key references s_order_no of sales_order.
Product_no	Varchar2	6	Primary key/foreign key refernces 'product_no of product_master
Qty_order	Number	8	
Qty_disp	Number	8	
Product_rate	Number	10.2	

Insert the following data into their respective tables using insert statement :

Data for sales_man master table :

Salesman No	Salesman Name	Address	City	Pin Code	State	Salamt	Tgt_to _get	Ytd	Remarks
500001	Kiran	A/14 worli	Bom bay	400002	Mah	3000	100	50	Good
500002	Manish	65,nari man	Bom bay	400001	Mah	3000	200	100	Good
500003	Ravi	P-7 Bandra	Bom bay	400032	Mah	3000	200	100	Good
500004	Ashish	A/5 Juhu	Bom bay	400044	Mah	3500	200	150	Good

(ii) **Data for Salesorder table :**

S_orderno	S_orderdate	Client no	Dely	Bill	Salesman	Delay	Orderstatus
019001	12-Jan-96	0001	F	N	50001	20-Jan-96	Ip
019002	25-Jan-96	0002	P	N	50002	27-Jan-96	C
016865	18-Feb-96	0003	F	Y	5000403	20-Feb-96	F
019003	03-Apr-96	0001	F	Y	500001	07-Apr-96	F
046866	20-May-96	0004	P	N	500002	22-May-96	C
010008	24-May-96	0005	F	N	500004	26-May-96	Ips

(iii) **Data for sales_order_details table :**

S_order no	Product no	Qty ordered	Qty disp	Product-rate
019001	P00001	4	4	525
019001	P07965	2	1	8400
019001	P07885	2	1	5250
019002	P00001	10	0	525
046856	P07868	3	3	3150
046865	P07885	10	10	5250
019003	P00001	4	4	1050
019003	P03453	2	2	1050
046866	P06734	1	1	12000
046866	P07965	1	0	8400
010008	P07975	1	0	1050
010008	P00001	10	5	525

Objective 4 : To implement the structure of the table :

Modifying the Structure of Tables : After table command is used to changing the structure of a table, Using the alter table clause you cannot perform the following tasks :

(*i*) change the name of table

(*ii*) change the name of column

(*iii*) drop a column

(*iv*) decrease the size of a table if table data exists.

The following tasks you can perform through alter table command.

(*i*) Adding new columns :

Syntax

ALTER TABLE tablename

ADD (newcolumname newdatatype (size));

(*ii*) Modifying existing table

Syntax

ALTER TABLE tablename

MODIFY (newcolumname newdatatype (size));

Note : Oracle not allow constraints defined using the alter table, if the data in the table, violates such constraints.

Removing/Deleting Tables : Following command is used for removing or deleting a table.

Syntax :

DROP TABLE tablename :

Defining Integrity constraints in the ALTER TABLE command

You can also define integrity constraints using the constraint clause in the ALTER TABLE command. The following examples show the definitions of several integrity constraints.

(1) **Add PRIMARY KEY**

Syntax

ALTER TABLE tablename

ADD PRIMARY KEY (columnname);

(2) **Add FOREIGN KEY**

Syntax :

ALTER TABLE tablename

ADD CONSTRAINT constraintname

FOREIGN KEY (columnname) REFERENCES tablename ;

Dropping integrity constraints in the ALTER TABLE command :

You can drop an integrity constraint if the rule that if enforces is no longer true or if the constraint is no longer needed. Drop the constraint using the ALTER TABLE command with the DROP clause. The following examples illustrates the dropping of integrity constraints.

(1) **DROP the PRIMARY KEY**

Syntax :

ALTER TABLE tablename

DROP PRIMARY KEY

(2) **DROP FOREIGN KEY**

Syntax :

ALTER TABLE tablename

DROP CONSTRAINT constraintname;

Objective 4 : To implement the concept of indexes and views.

Indexes : An index is an ordered list of content of a column or group of columns in a table. An index created on the single column of the table is called simple index. When multiple table columns are included in the index it is called composite index.

Creating an Index for a table :

Syntax (Simple)

CREATE INDEX index_name

ON tablename (column name);

Composite Index :

CREATE INDEX index_name

ON tablename (columnname, columnname);

Creating an UniQuestion Index :

CREATE UNIQUESTION INDEX indexfilename

ON tablename(columnname);

Dropping Indexes :

An index can be dropped by using DROP INDEX

Syntax :

DROP INDEX Index filename;

Views : Logical data is how we want to see the current data in our database. Physical data is how this data is actually placed in our database.

Views are masks placed upon tales, This allows the programmer to develop a method via which we can display predetermined data to users according to our desire.

Views may be created fore the following reasons :

1. The DBA stores the views as a definition only. Hence there is no duplication of data.
2. Simplifies Questionries.
3. Can be Questionries as a base table itself.
4. Provides data security.
5. A voids data redundancy.

Creation of Views :

Syntax :

```
CREATE VIEW viewname AS
SELECT columnname, columnname
FROM tablename
WHERE columnname=expression_list;
```

Question 1. Create the following tables :

Challan Header

Column Name	**Data**	**Size**	**Attributes**
Challan_no	varchar2	6	Primary key
S_order_no	varchar2	6	Foreign key reference s_order_no
of sales_order table			
challan_date	date		not null
billed_yn	char	1	values ('Y', 'N'), Default 'N'

Table Name : Challan_Details

Column Name	**Data**	**Size**	**Attributes**
Challan_no	varchar2	6	Primary key/Foreign key
references			
Product_no of product_master			
Qty_disp	number	4.2	not null

Q.2 Insert the following values into the challan header and challan_details tables :

(*i*)

Challan No	**S_order No**	**Challan Date**	**Billed**
CH9001	019001	12-DEC-95	Y
CH865	046865	12-NOV-95	Y
CH3965	010008	12-OCT-95	Y

Data for challan_details table

Challan No	**Product No**	**Qty Disp**
CH9001	P00001	4
CH9001	P07965	1
CH9001	P07885	1
CH6865	P07868	3
CH6865	P03453	4
CH6865	P00001	10
CH3965	P00001	5
CH3965	P07975	2

Objective : Answer the following Questionries :

Q.1 Make the primary key to client_no in client_master.

Q.2 Add a new column phone_no in the client_master table.

Q.3 Add the not null constraint in the product_master table with the columns description, profit percent, sell price and cost price.

Q.4 Change the size of client_no field in the client_master table.

Q.5 Select product_no, description where profit percent is between 20 and 30 both inclusive.

Objective 5 : To implement the concept of Joins

Joint Multiple Table (Equi Join) : Some times we require to treat more than one table as though manipulate data from all the tables as though the tables were not separate object but one single entity. To achieve this we have to join tables. Tables are joined on column that have dame data type and data with in tables.

The tables that have to be joined are specified in the FROM clause and the joining attributes in the WHERE clause.

Algorithm for JOIN in SQL :

1. Cartesian product of tables (specified in the FROM clause)
2. Selection of rows that match (predicate in the WHERE clause)
3. Project column specified in the SELECT clause.

1. CARTESIAN PRODUCT :

Consider two table student and course

```
Select B.*, P.*
FROM student B, course P;
```

2. INNER JOIN :

Cartesian product followed by selection

```
Select B. *, P.*
FROM student B, Course P
WHERE B.course # P.course # ;
```

3. LEFT OUTER JOIN :

LEFT OUTER JOIN = Cartesian product + selection include rows from the left table which are unmatched pat nulls in the values of attributes belonging to the second table

Exam :

```
Select B.*, P.*
FROM student B left join course P
ON B.course # P.course #;
```

4. RIGHT OUTER JOIN :

RIGHT OUTER JOIN = Cartesian product + selection but include rows form right table which are unmatched.

Exam :

Select B.*, P .*

From student B RIGHT JOIN course P

B.course # = P course # ;

5. FULL OUTER JOIN

Exam

Select B.*, P.*

From student B FULL JOIN course P

On B.course # = P course #;

OBJECTIVE : Answer the following Questionnaires :

Questions :

1. Find out the product which has been sold to 'Ivan Sayross'.
2. Find out the product and their quantities that will have do delivered.
3. Find the product_no and description of moving products.
4. Find out the names of clients who have purchased 'CD DRIVE'.
5. List the product_no and s_order_no of customers having qty ordered less than 5 from the order details table for the product "1.44 floppies".
6. Find the products and their quantities for the orders placed by 'Vandand Saitwarl' and "Ivan Bayross".
7. Find the products and their quantities for the orders placed by client_no "C0001" and "C00002".
8. Find the order No., Client No and salesman No., where a client has been received by more than one salesman.
9. Display the s_order_date in the format "dd-mm-yy" e.g., "12-Feb-96".
10. Find the date, 15 days after date.

Objective 6 : To implement the concept of grouping of Data.

Grouping Data From Tables : There are circumstances where we would like to apply the aggregate function not only to a single set of tuples, but also to a group of sets of tuplets, we specify this wish in SQL using the group by clause. The attributes or Attributes given in the group by clause are used to form group. Tuples with the same value on all attributes in the group by clause are placed in one group.

Syntax :

```
SELECT columnname, columnname
FROM tablename
GROUP BY columnname;
```

At times it is useful to state a condition that applies to groups rather than to tuples. For example we might be interested in only those branches where the average account balance is more than 1200. This condition does not apply to a single tuple, rather it applies to each group constructed by the GROUP BY clause. To express such Questionary, we use the having clause of SQL. SQL applies predicates in the having may be used.

Syntax :

```
SELECT columnname, columnname
FROM tablename
GROUP BY columnname;
HAVING searchcondition;
```

Objective : Answer the following Questionries :

Q.1. Print the description and total quantity sold for each product.

Q.2. Find the value of each product sold.

Q.3. Calculate the average quantity sold for each client that has a maximum order value of 15000.

Q.4. Find out the products which has been sold to Ivan.

Q.5. Find the names of clients who have 'CD Drive'.

Q.6. Find the products and their quantities for the orders placed by 'Vandana' and 'Ivan'.

Q.7. Select product_no, total qty_ordered for each product.

Q.8. Select product_no, product description and qty ordered for each product.

Q.9. Display the order number and dry on which clients placed their order.

Q.10. Display the month and Date when the order must be delivered.

Objective 7 : To implement the concept of SubQuestionries.

SubQuestionries : A subQuestionry is a form of an SQL statement that appears inside another SQL statement. It also termed as nested Questionary. The statement containing a subQuestionry called a parent statement. The rows returned but the subQuestionary are use by the following statement.

It can be used by the following commands :

1. To insert records in the target table.
2. To create tables and insert records in this table.
3. To update records in the target table.
4. To create view.
5. To provide values for the condition in the WHERE, HAVING IN, SELECT, UPDATE, and DELETE statements.

EXAMPLE :

Creating clientmaster table from oldclient_master, table

```
Create table client_master
AS SELECT * FROM oldclient_master;
```

Using the Union, Intersect and Minus Clause ;

Union Clause : The user can pu together multiple Questionnaires and combine their output using the union clause. The union clause merges the output of two or more Questionnaires into a single set of rows and column. The final output of union clause will be

Output : = Records only in Questionary one + records only in Questionary two + A single set of records with is common in the both Questionries.

Syntax :

```
SELECT columnname, columnname
FROM tablename1
UNION
SELECT columnname, columnname
From tablename2;
```

Intersect Clause : The use cane put together multiple Questionnaires and their output using the interset clause. The final output of the intereset clause will be :

Output = A single set of records which are common in both Questionries.

Syntax :

```
SELECT columnname, columnname
FROM tablename1
INTERSECT
SELECT columnname, columnname
FROM tablename2;
```

OBJECTIVE 8: To implement the concept of Index and views

Indexes : An index is an ordered list of content of a column or group of columns in a table. An index created on the single column of the table is called simple index. When multiple table columns are included in the index it is called composite index.

Creating an index for a table :

Syntex (Simple)

```
CREATEINDEX index_name
ON tablename (column name)
```

Composite Index :

```
CREATEINDEX index_name
ON tablename (columnname,columnname);
```

Creating an UniQuestion Index :

```
CREATE UNIQUESTION INDEX indexfilename
ON tablename (columnname);
```

DROPPING INDEXES :

An index can be dropped by using DROP INDEX

SYNTAX :

```
DROP INDEX indexfilename;
```

Views : Logical data is how we want to see the current data in our database. Physical data is how this data ius actually placed in our database.

Views are masks placed upon tables. This allows the programmer to develop a method via which we can display predetermined data to users according to our desire.

Views may be created fore the following reasons:

1. The DBA stores the views as a definition only. Hence there is no duplication of data.
2. Simplifies Questionries.
3. Can be questionried as a base table itself.
4. Provides data security.
5. Avoids data redundancy.

Creation of Views :

Syntex :

```
CREATE VIEW viewname AS
SELECT columnname,columnname
FROM tablename
WHERE columnname = expression_list;
```

Renaming the columns of a view:

Syntex :

```
CREATE VIEW viewname AS
SELECT newcolumnname.....
FROM tablename
WHERE columnname = expression_list
```

Selecting a data set from a view:

Syntex :

```
SELECT columnname, columnname
FROM viewname
WHERE search condition;
```

Destroying a view:

Syntex :

```
DROP VIEW viewname
```

Objective : Answer the following questions :

Q.1. Create an index on the table client_master, field client_no.

Q.2. Create an index on the sales_order, field s_order_no.

Q.3. Create an composite index on the sales_order_details table for the columns s_order_no and product_no.

Q.4. Create an composite index ch_index on challan_header table for the columns challan no and s_order_no.

Q.5. Create an uniQuestion index on the table salesman_master, field salesman_no.

Q.6. Drop index ch_index on the table challan_header.

Q.7. Create a view client_view on client_master and rename the columns as name, add1, add2, city,pcode, state respectively.

Q.9. Select the client names from client_view who lives in city 'Bombay'

Q.10. Drop the view client_view.

Objective 9 : To implement the basics of PL/SQL.

Introduction : PL/SQL bridge the gap between database technology and procedural programming languages. It can be thought of as a development tool extends the facilities of Oracles SQL database language. Via PL/SQL you can insert, delete, update and retrieve table as well as use procedural techniquestations such as writing loops or branching to another block of code.

PL/SQL Block Structure

```
DECLARE
Declaration of memory variables used later
BEGINS
SQL executable statements for manipulating table data>
EXCEPTIONS
SQL and/or PL.SQL code to handle errors.
END
```

Displaying user Messages on the screen : Any programming tool requires a method through which messages can be displayed to the user.

dbms output is a package that includes a number of procedure and functions that accumulate information in a buffer so that it can be retrieved later. These functions can also be used to display messages to the user.

Put line : Put a piece of information in the buffer followed by a end of line marker. It can setting the server output on :

SET SERVER OUTPUT ON :

Example : Write the following code in the PL/SQL block to display messages to user DBMS_OUTPUT.PUT_LINE ('Display user message')

Conditional control in PL/SQL -

Syntex :

```
IF <condition>THEN
<Action>
ELSEIF <condition>
<Action>
ELSE
<Action>
ENDIF;
```

The WHILE LOOP

Syntex :

```
WHILE <condition>
LOOP
<Action>
END LOOP;
```

The FOR LOOP statement :

Syntex :

```
For variable IN [REVERSE] start – end
LOOP
<Action>
END LOOP;
```

The GOTO Statement : The goto statement allows you to change the change the flow of control within a PL/SQL Block.

Q.1 WAP in PL/SQL for addition of two numbers.

Q.2 WAP in PL/SQL for addition 01 to 100 numbers.

Q.3 WAP in PL/SQL to check the given number is even or odd.

Q.4 WAP in PL/SQL to inverse a number, eg., Number 5639 when inverter must be display.

Q.5 WAP in PL/SQL for changing the price of product 'P00001' to 4000 if the price is low than 4000 in product_master table. The change is recorded in the old_price_table along with product_no and the date on which the price was changed last.

Objective 10 : To implement the concept of Cursor and Trigger.

Cursor : We have seen how oracle executes an SQL statement. Oracle DBA uses a work area for its internal processing. This work are is private to SQL's operation and is called a cursor.

The data that is stored in the cursor is called the **Active Data Set**. The size of the cursor in memory is the size required to hold the number of rows in the Active Data Set.

Explicit Cursor : You can explicitly declare a cursor to process the rows individually. A cursor declared by the user is called **Explicit Cursor**. For Questionries that return more than one row, You must declare a cursor explicitly.

The data that is stored in the cursor is called the **Active Data Set**. The size of the cursor in memory is the size required to hold the number of rows in the Active.

Why use an Explicit Cursor : Cursor can be used when the user wants to process data one row at a time.

Explicitly Cursor Management : The steps involved in declaring a cursor and manipulating data in the active data set are :

- Declare a cursor that specifies the SQL select statement that you want to process.

- Open the Cursor.
- Fetch the data from the cursor one row at a time.
- Close the cursor.

Explicit Cursor Attributes : Oracle provides certain attributes/cursor variables to control the execution of the cursor. Whenever any cursor (explicit or implicit) is opened and used Oracle creates a set of four system variables via which Oracle keeps track of the 'Current' status of the cursor. You

- Declare a cursor that specifies the SQL select statement that you want to process.
- Open the Cursor.
- Fetch the data from the cursor one row at a time.
- Close the cursor.

How to Declare the Cursor :

The General Syntax to create any particular cursor is as follows :

Cursor <Cursorname> is Sql Statement;

How to Open the Cursor

The General Syntax to Open any particular cursor is as follows :

Open Cursorname;

Fetching a record From the Cursor :

The fetch statement retrieves the rows from the active set to the variables one at a time. Each time a fetch is executed. The focus of the DBA cursor advances to the next row in the Active set.

One can make use of any loop structure (Loop-End Loop along with While, For) to fetch the records from the cursor into variable one row at a time.

The General Syntax to Fetch the records from the cursor is as follows :

Fetch cursorname into variable1, variable2, ...

Closing a Cursor :

The General Syntax to Close the cursor is as follows :

Close <cursorname>;

Database Triggers :

Database triggers are procedures that are stored in the database and are implicitly executed (fired) when the contents of a table are changed.

Use of Database Triggers :

Database triggers support Oracle to provide a highly customized database management system. Some of the uses to which the database triggers can be put to customize management information in Oracle are as follows :

- A Trigger can permit DML statements against a table only if they are issued, during regular business hours or on predetermined weekdays.
- A trigger can also be used to keep an audit trail of a table along with the operation performed and the time on which the operation was performed.

- It can be used to prevent invalid transactions.
- Enforce complex security authorizations.

How to Apply Database Triggers :

A trigger has three basic parts :

1. A triggering event of statement.
2. A trigger restriction.
3. A trigger action.

Type of Triggers :

Using the various options, four types of triggers can be created :

1. **Before Statement Trigger :** Before executing the triggering statement, the trigger action is executed.
2. **Before Row Trigger :** Before modifying the each row affected by the triggering statement and before appropriate integrity constraints, the trigger is executed if the trigger restriction either evaluated to TRUE or was not included'.
3. **After Statement Trigger :** After executing the triggering statement and applying ay deferred integrity constraints, the trigger action is executed.
4. **After Row Trigger :** After modifying each row affected by the triggering statement and possibly applying appropriate integrity constraints, the trigger action is executed for the current row if the trigger restriction either evaluates to TRUE or was not included.

Syntax For Creating Trigger

The syntax for Creating the Trigger is as follows :

```
Create or replace Trigger<Triggername> (Before, After) (Delete, Insert, Update) On
<Tablename> For Each row when Condition
Declare
<Variable declarations>;
<Constant Declarations>;
Begin
<PL/SQL> Subprogram Body;
Exception
Exception Pl/SQL block;
End;
```

How to Delete a Trigger :

The syntax for Deleting the Trigger is as follows :

DROP Trigger <Triggername>;

Q.1 Two Tables are there

Table Name : 0-Employee

Column Name	Data	Size	Attributes
Emp_Code	Varchar2	10	Primary Key
Ename	Varchar2	20	The name of the Candidate
Deptno	Number	5	The Department No
Job	Varchar2	20	The Employee Job Detail
Salary	Number	8.2	The Current salary of Em

Table Name : Emp_Employee

Column Name	Data	Size	Attributes
Emp_Code	Varchar2	10	Primary Key
Raise_Date	date	20	The Date on which the raise was given
Raise_Amt	Number	8.2	The raise given to the employee

The HRD manager has decided to raise the salary for all the employees in the Dept No 20 by 5%. Whenever any such raise is given to the employees, a record for the same is maintained in the emp_raise table. It include the Emp No, the date when the raise was given and the actual raise. Write a PL/SQL block to update the salary of each employee and insert a record in the emp_raise table.

Q.2 Two Tables are there

Table Name : Client_Master

Column Name	Data Type	Size	Attributes
Clinet_No	Varchar2	6	Primary Key/First letter must start with 'C'
Name	Varchar2	20	No Null
City	Varchar2	10	
State	Varchar2	10	
Bal_Due	Number	10,2	

Table Name : auditclient

Column Name	Data Type	Size	Attributes
Clinet_No	Varchar2	10	Primary Key
Name	Varchar2	20	
Bal_Due	Number	10,2	
Operation	Varchar2	8	
Odate	Date		

Create a transparent audit system for a table Client_master. The system must keep track of the records that are being detected or modified and when they have been deleted or modified.

Objective 11 : To implement the concept of Forms and reports.

The object of the lab is to create a form using a parent table and a chile table to take advantage of the schema's relationships.

A data block in Oracle Forms = A table in the database

Each block contains items which equal table columns or fields.

These are arranged into records.

1. Start Schema Builder. Open S_Customer and S_Order or S_Order1.
2. Start Form Builder. use the data block wizard to create a form for S_Customer, including the Name,
3. After the form is created, click on Window on the Object Navigator to expand it. Right click on Window1. Click on Property Pallet. Go to Physical on property pallet. Make sure Show Horizontal Scroll Bar and Show Vertical Scroll Bar both are YES.
4. Run the form. Execute the Questionary. Notice that data already exists in this table.
5. Highlight Data Blocks in the Object Navigator. Go up to Tools – Data Block Wizard.
6. Create a form for S_Order or S_Order 1.
7. Include the columns ID – Customer_ID – Date_Ordered – Date_Shipped – Total.
8. Click Create Relationship. Click OK. Make sure Autojoin Datablocks is checked.
9. Check Detail item to Customer_ID and Master Item to ID. This says that the parent table, the table on the one side of the relationship has the primary key of ID in the S_Customer table, and the foreign key on the many side is Customer_ID in the S_Order table. This relationship can be seen if you open schema builder and look at the tables and the relationship between them.
10. Make the layout tabular.
11. Records displayed will be 5 and Display Scrollbar will be checked off.
12. Run the form and execute the Questionary. Scroll through the data and notice that the orders are linked with the customers.
13. If you input a detail, the foreign key is automatically filled with the value of the current primary key displayed by the customer.

To delete a record, you must first delete the details, save the change, then delete the master and save. After you make any change, save the data. If there is any problem with integrity, the record won't save.

If you have two table joined by a relationship table, the relationship table MUST contain valid data or you will not be able to add data to the detail table.

IE – If you have two tables and one relation table such as – Students – Have – Marks.

The have table would include at least one column, a student ID would have a exist in the Have table, and a valid Mark ID must exist in the marks table that = a StudentID in the Have table which = a StudentID in the Student table.

To create a grouped report :

1. Start report builder.
2. Select "use the report wizard"
3. Title the report "Maintenance Bill" and select Group Above and click next.
4. Click on Questionry Builder.
5. Select the tables Jdrivers, JmaintenanceBill, JMBCCodeand costs, and Jvendors.
6. Select the fields by double clicking them. A check mark should appear to the left of the field when it is selected. The fields you should select are :
   - Jdrivers – Drsupervisor
   - JmaintenanceBill – All field except VandorNumber
   - JMBC Codeandcosts – All except Mbillnumber
   - Jvendors – Vencompanyname
7. Click OK, Click Next.
8. The grouping window will now open. Select Mbillno and include it. Above it on the right hand side, the words Group 1 should appear. Next select Vencompanyname. The words Group 2 should appear above it. Click on the Vencompanyname you have just selected on the right and drag it up to group 1. It is now included in group 1 if the Group 2 is no longer displayed. Do this for the remaining fields. Include all fiends EXCEPT SRVCODE, SCPARTCOST, and SCLABCOST, Click next.
9. Now you can include the remaining fields you left out before. Ie-SRVCODE, SCPARTCOST, and SCLABCOST, Click next.
10. We will do two sums on some fields. Select SCPARTCOST and click the sum button. Do the same for SCLABCOST. Click next.
11. When choosing the layout template, select Cyan Grid. Click finished.
12. If the layout is not as you would like, you can edit the layout by clicking on View on the menu bar and selecting Layout Model.

❑❑❑

COMPETITIVE QUIZ

COMPETATIVE QUIZ

SHORT ANSWER TYPE QUESTIONS

1. Explain the difference between a database administrator and a data administrator.

Ans. Database Administrator : A person (or group of people) responsible for the maintenance and performance of a database and responsible for the planning, implementation, configuration, and administration of relational database management systems.

Data Administrator : The individual or organization responsible for the specification, acquisition, and maintenance of data management software and the design, validation, and security of files or databases. The DA is in charge of the data dictionary and data model.

2. Explain the difference between an explicit and an implict lock.

Ans. Explicit Lock : Lock is explicity requested for a record or table.

Implict Lock : Lock is implied but is not acquired.

3. What is lock granularity ?

Ans. There are many locks available for the database system to have like Intent Shared, Shared, Intent exclusive, exclusive and shared Intent exclusive. Locking granularity refers to the size and hence the number of locks used to ensure tehe consistency of a database during multiple concurrent updates.

4. In general, how should the boundaries of a transaction be defined ?

Ans. A Tranaction ensures that one or more operations execute as an atomic unit of work. If one of the operations with in a transaction fails, then all of them are rolled-back so that the application is returned to its prior state. The boundaries that define a group of operations done within a single transaction.

5. Explain the meaning iof the expression ACID transaction.

Ans. ACID means Atomic, Consistency, Isolation, Durability, so when any transaction happen it should be Atomic that is it should either be complete or fully imcomplete. There should not be anything like Semi complete. The Database State should remain consistent after the completion of the transaction. If there are more than one Transaction then the transaction should be scheduled in such a fashion that they remain in Isolation of one another. Durability means that Once a transaction commits, its effects will persist even if there are system failures.

6. Explain the necessity of defining processing rights and responsibilities. How are such responsibilities enforred ?

Ans. One of the reason to define rights is the security in the database system. If any user is allowed to define the data or alter the data then the database would just be no use and so processing rights and responsibilities are clearly defined in any database system. The responsibilties are enforced using the table space provided by the database system.

7. Describe the advantages and disadvantages of DBMS-provided and application-provided security.

Ans. DBMS provided security : Any database system requires you to login and then process the data depending on the rights given by the DBA to the user who has logged in. The advantage of such a system is securing the data and providing the user and the DBA the secured platform. Any user who logs is cannot do whatever he want but his role can be defined very easily. There is no major disadvantage about the DBMS provided security apart from overahead of storing the rights and priviledges about the users.

Application-provided secutiry : It is much similar to the DBMS provided security but the only difference is that its the duty of programmer creating the application to provide all the securities so that the data is not mishandled.

8. Explain how a database could be recovered via reprocessing. Why is this generally not feasible ?

Ans. If we reprocess the transaction then the database can be made to come to a state where the database is consistent and so reprocessing the log can recover the database. Reprocessing is not very feasible for a very simple reason that its very costly from time point of view and requires lots of rework and many transaction are even rollback giving more and more rework.

9. Define rollback and rollforward.

Ans. Rollback : Undoing the changes made by a transaction before it commits or to canel any changes to a database made during the current transaction.

Rollforward : Re-doing the changes made by a transaction after it commits or to overwrite the changed value again to ensure consistency.

10. Why is it important to write to the log before changing the database values?

Ans. The most important objective to write the log before the database is changed is if there is any need to rollback or rollforward any transaction then if the rollback rollforward cannot be done accurately.

MULTIPLE CHOICE QUESTIONS

1. In a relationship .... represents the maximum numbers of records in one entry that are linked to a single now in another entity.

(*a*) Degree (*b*) Cardinality
(*c*) Domain (*d*) Link

2. A compact disc provides .... type of data access.

(*a*) Sequential (*b*) Direct
(*c*) Both (*a*) and (*b*) (*d*) None of these

3. Data Integrity refers to :

(*a*) Non-duplication of data

(*b*) Accuracy of data
(*c*) Security of data
(*d*) Centralised data

4. Which of the following would be considered records ?
(*a*) Drawings (*b*) Microfilm
(*c*) Routing slips (*d*) Motion Pictures

5. Which is not a component of a relational database ?
(*a*) Entity (*b*) Attribute
(*c*) Table (*d*) Hierarchy

6. A (n) .... is a detailed technology independent specification of the overall structure of the database ?
(*a*) User view
(*b*) Physical schema
(*c*) Conceptual scheme
(*d*) External schema

7. Which is Not a disadvantage of traditional File Processing Systems ?
(*a*) Program-data dependence
(*b*) Reduced data redundancy
(*c*) Laughty development time
(*d*) Limited data sharing

8. In .... normal forms, any multi-valued attributes have been removed.
(*a*) First (*b*) Second
(*c*) Third (*d*) Fourth

9. A (*n*) .... is an attribute in a relation that serves as a primary key of another relation in the same.
(*a*) Composite key (*b*) Foreign key
(*c*) Identifier key (*d*) Primary key

10. Data is the ...., that describe people, exists, things and ideas.
(*a*) Words (*b*) Graphics
(*c*) Numbers (*d*) All of these

11. .... File organisation rewrites the file to be rewritten when new records are added.
(*a*) Sequential (*b*) Direct
(*c*) Indexed (*d*) Hashed

12. A (*n*) .... data base is one where data is stored in a collection of tables.
(*a*) Hierarchical (*b*) Direct
(*c*) Relational (*d*) Object-oriented

13. Father and son-share .... type of relationship.
(*a*) 1 : μ (*b*) μ : μ
(*c*) 1 : 1 (*d*) None of these

14. The database model that uses a series of two-dimensional tables or files to store information is :
(*a*) Object-oriented database model
(*b*) Hierarchical database
(*c*) Relational database model
(*d*) Network database model

15. A logical collection of information that supports business analysis activities and decision-making tasks is :
(*a*) An object-oriented database
(*b*) A data warehouse
(*c*) A database
(*d*) A detaining tool

16. You want to run a SELECT command that lists all employees by the ascending order of their names. What clause would you include in the query to do the ordering ?
(*a*) List ASC BY employee name
(*b*) ORDER By employee name
(*c*) SORT BY employee name
(*d*) WHERE employee name ASC

17. A language that is used to specify the database scheme through a set of definition is called :
(*a*) SQL (*b*) DmL
(*c*) DPL (*d*) ISO

18. Which of the following is an aggregate function ?
(*a*) PLUS (*b*) ABS
(*c*) Minimum (*d*) Count

19. SQL is .... type of language.
(*a*) Low level (*b*) Procedural
(*c*) High level (*d*) Non-procedural

20. Pre query : SELECT * FROM Student WHERE Course LIKE '%' will?
(*a*) Display all the courses which begins with the latter '*a*'
(*b*) Display all the courses which ends with the latter '*a*'
(*c*) Display only those courses which contains numeric digits in the beginning
(*d*) Produce on error because the query is incorrect

21. What does SQL stand for ?
(*a*) Structured Question language
(*b*) Structured Query language
(*c*) Strong Query language
(*d*) Structured Quarantine language

22. Which SQL command insects new data is a database ?
(*a*) INSERT (*b*) NEW
(*c*) ENTER (*d*) ADD

23. .... Command extracts data from a database ?
(*a*) OPEN (*b*) EXTRACT
(*c*) GET (*d*) SELECT

24. Which SQL keyward is used to sort the result ?
(*a*) SORT (*b*) ORGANISE
(*c*) ORGANISE (*d*) ARRANGE

25. How may rules are there is E.F. Codd's Rule ?
(*a*) 10 (*b*) 12
(*c*) 8 (*d*) 9

26. In a relational schema, each tuple is divided into fields called :
(*a*) Relations (*b*) Domains
(*c*) Queries (*d*) All of the above

27. A top-to-bottom relationship among the items in a database is established by a :
(*a*) Hierarchial schema
(*b*) Network schema
(*c*) Relational schema
(*d*) All of the above
(*e*) None of the above.

28. A good query system :
(*a*) can accept English language commands
(*b*) allows non-programmers to access information stored in a database
(*c*) can be accessed only by data processing professionals
(*d*) Both (*a*) and (*b*)
(*e*) None of the above.

29. A logical schema
(*a*) is the entire database
(*b*) is a standard way of organizing information into accessible parts
(*c*) describes how data is actually stired on disk
(*d*) All of the above
(*e*) None of the above.

30. A report form :
(*a*) appears on the computer monitor during data entry
(*b*) is used during report generations to format data
(*c*) Both (*a*) and (*b*)
(*d*) all of the above (*e*) None of the abov

31. A command that lets you change one or more fields in a record is :
(*a*) Insert (*b*) Modify
(*c*) Look-up (*d*) All of the above
(*e*) None of the above

32. The modify operation is likely to be done after :
(*a*) Delete (*b*) Look up
(*c*) Insert (*d*) All of the above
(*e*) None of the above

33. The database environment has all of the following components expert :
(*a*) users (*b*) separate files
(*c*) database
(*d*) database administrator
(*e*) None of the above

34. Which of the following is not characteristic of a relational database model ?
(*a*) tables
(*b*) treelike structures

(c) complex logical relationships
(d) records

35. Data Security threats include :
(a) hardware failure
(b) privacy invasion
(c) fraluent manipulation of data
(d) All of the above
(e) None of the above

36. Which is the abbreviation used for a sofware package that permits the user to create, retrieve and maintain records in a database ?
(a) DASD (b) Fms
(c) EMMS (d) DBMS
(e) None of the above

37. Which of the following is not a relational database ?
(a) dBASE IV (b) 4th Dimension
(c) FoxPro (d) Reflex
(e) None of the above

38. Which language has recently become the defect standard for interfacing application program with relation database system ?
(a) Oracle (b) SQL
(c) DBASE (d) 4Gl
(e) None of the above

39. A scheme describe :
(a) data elements (b) records and files
(c) record relationships
(d) All of the above
(e) None of the above

40. An audit trail :
(a) is used to make back-up copies
(b) is the recorded history of operations performed on a file
(c) can be used to restore lost information
(d) All of the above
(e) None of the above

41. If a piece of data is stored in two places in the database, then :
(a) storage space in wasted
(b) changing the data is one spot will cause data inconsistency
(c) it can be more easily accessed
(d) both (a) and (b)
(e) None of the above

42. In order to use a DBMS, it is important to understand
(a) the physical schema
(b) all substances that the system supports
(c) One subscheme
(d) both (a) and (b)
(e) None of the above

43. The index consits of :
(a) a list of keys
(b) pointers to the master list
(c) both (a) and (b)
(d) All of the above
(e) None of the above.

44. Subschema can be used to :
(a) cretae very different, personalized view of the same data
(b) persent information in different formats
(c) hide sensitive information by omitting fields from the sub-schema's description
(d) All of the above
(e) None of the above

45. A data dictionary doesn't provide information about
(a) where data is located
(b) the size of the disk steerage device
(c) who owns or is responsible for the data
(d) how the data is used
(e) security and privacy limitations

46. Data items grouped together for storage purposes are called a :
(a) record (b) title
(c) list (d) string
(e) None of the above

47. Periodically adding, changing and deleting file records is called file
(a) Updating (b) upgrading
(c) restructing (d) renewing
(e) None of the above

48. What name is given to the collection of facts, items of information or data which are related in some way ?
(*a*) Database
(*b*) Directory information
(*c*) Information tree
(*d*) Information provider
(*e*) None of the above.

49. A computer file contains several records. What does each record contains ?
(*a*) Bytes (*b*) Words
(*c*) Fields (*d*) Database
(*e*) None of the above.

50. The way a particular application views the data from the database that the application uses is a :
(*a*) module (*b*) relational model
(*c*) schema (*d*) subschema
(*e*) None of the above.

51. What is the name given to the database management system which is able to handle full text data, image data, audio and idea ?
(*a*) Full media (*b*) Graphics media
(*c*) Multimedia (*d*) Hypertext
(*e*) None of the above

52. In any hierarchy of data organization, the smallest entity to be processed as a single unit is called :
(*a*) data field (*b*) data record
(*c*) data file (*d*) database
(*e*) None of the above.

53. A condition that led to the developmemnt of database was :
(*a*) a demand for more data to support information needs
(*b*) an increase in the amount of data handles by organizations
(*c*) the proliferation of data files
(*d*) All of the above
(*e*) None of the above

54. A report generatory is used to :
(*a*) update files
(*b*) print files on paper
(*c*) data entry (*d*) All of the above
(*e*) None of the above

55. Two files may be joined into a third file if :
(*a*) they have a row in common
(*b*) they have a field in common
(*c*) they have no records with the same value in the common field
(*d*) both (*b*) and (*c*)
(*e*) None of the above

56. A network schema
(*a*) restricts the structure to a one-to-many relationship :
(*b*) permitsmany-to-many relationships
(*c*) stores data is tables
(*d*) All of the above
(*e*) None of the above

57. In alarge DBMS :
(*a*) each user cam "see" only a small part of the entire database
(*b*) each user can access every subschema
(*c*) each subschema contains every field in the logical schema
(*d*) All of the above
(*e*) None of the above

58. An indexing operation
(*a*) sorts a file using a single key
(*b*) sorts file using two keys
(*c*) establishes an index for a file
(*d*) both (*b*) and (*c*)
(*e*) None of the above

59. What is the language used by most of the DBMS for helping heir users to access data ?
(*a*) High level language
(*b*) Query language
(*c*) SQL
(*d*) 4GL
(*e*) None of the above

60. A relational database management (RDBMS) package manages data in more than one files as once. How does it organize these file ? As :

(*a*) tables (*b*) relations
(*c*) tuple (*d*) both (*a*) and (*b*)
(*e*) None of the above

61. The distinguishable parts of a record are called :
(*a*) files (*b*) data
(*c*) fields (*d*) database
(*e*) None of the above

62. A number of related records that are treated as a unit called
(*a*) file (*b*) field
(*c*) data (*d*) batch
(*e*) None of the above

63. Large collections of files are called :
(*a*) fields (*b*) records
(*c*) database (*d*) file system
(*e*) None of the above

64. The database administration function includes :
(*a*) application programming
(*b*) computer operations management
(*c*) database access planning
(*d*) All of the above
(*e*) None of the above

65. Updating a database means
(*a*) revising the file structure
(*b*) recorganizing the database
(*c*) modifying or adding record occurrences
(*d*) All of the above
(*e*) None of the above

66. Which two files are used during operation of the DBMS?
(*a*) query language and utilities
(*b*) data manipulation language and query language
(*c*) data dictionary and transaction log
(*d*) data dictionary and query language
(*e*) None of the above

67. A locked file can be :
(*a*) accessed by only one user
(*b*) modified by users with the correct password
(*c*) is used to hid sensitive information
(*d*) both (*a*) and (*b*)
(*e*) None of the above

68. The PROJECT command will create new table that has :
(*a*) more fields than the original table
(*b*) more rows than the original table
(*c*) both (*a*) and (*c*)
(*d*) all the field of master table
(*e*) None of the above

69. A transparent DBMS :
(*a*) cannot hide sensitive information form users
(*b*) keep its logical structure hidden from users
(*c*) keeps its physical structure hidden from users
(*d*) both (*b*) and (*c*)
(*e*) None of the above

70. A database management system :
(*a*) allows simultaneous access to multiple files
(*b*) can do more than a record management system
(*c*) is a collection of program for managing data in a single file
(*d*) both (*a*) and (*b*)
(*e*) None of the above

71. A large computer information system maintain many different computer files. Which amongst them is called a perpetusal file ?
(*a*) Specialized file (*b*) Log file
(*c*) Master file (*d*) Update file
(*e*) None of the above

72. When using a database management system, the first thing that you must do is to :
(*a*) create a database file
(*b*) activates file editor
(*c*) load the software into your microcomputer
(*d*) keep a floppy disk in readiness
(*e*) None of the above

73. Related field in a data base are grouped to form :
(*a*) data file (*b*) data record
(*c*) menu (*d*) bank
(*e*) None of the above

74. A set of programs that handle a firm's database responsibility is called a
(*a*) Database Management System (DBMS)
(*b*) Database Processing System (DBMS)
(*c*) Data Management System (DBMS)
(*d*) All of the above
(*e*) None of the above

75. In the DBMS approach, application programs perform the :
(*a*) storage function
(*b*) processing functions
(*c*) access control
(*d*) All of the above
(*e*) None of the above

76. Queries to a database :
(*a*) are written in English
(*b*) can use aggregate functions like SUM and COUNT
(*c*) both (*a*) and (*b*)
(*d*) All of the above
(*e*) None of the above.

77. A form can be used to :
(*a*) modify records
(*b*) delete records
(*c*) format printed output
(*d*) All of the above
(*e*) None of the above

78. A record management system
(*a*) can handle many files of information at a time
(*b*) can be used to extract information stored in a computer file
(*c*) always uses a list as its model
(*d*) both (*a*) and (*b*)
(*e*) None of the above

79. The files stored on a secondary stage device are composed of a hierarchy of data. What does a record in a file contain ?
(*a*) Bits (*b*) Characters
(*c*) Data filed (*d*) Schema
(*e*) None of the above

80. The highest level in the hierarchy of data organization is called :
(*a*) data bank (*b*) database
(*c*) data file (*d*) data record
(*e*) None of the above

81. The language used in application programs to request data from DBMS is referred to as the
(*a*) DML (*b*) DDL
(*c*) query language (*d*) All of the above
(*e*) None of the above

82. Batch processing is appropriate if :
(*a*) a large computer system is available
(*b*) only a small computer system is available
(*c*) only a few transactions are involved
(*d*) All of the above
(*e*) None of the above

83. One approach to standardizing storing of data :
(*a*) MIS
(*b*) structured programming
(*c*) CODASYL specifications
(*d*) All of the above
(*e*) None of the above

84. Which of the following is a database administrate's function ?
(*a*) Database design
(*b*) Backing up the database
(*c*) Performance monitoring
(*d*) User coordination
(*e*) None of the above

85. Which of the following is a type of DBMS software ?
(*a*) Database manipulation language
(*b*) Query language
(*c*) Utilities
(*d*) Report writer
(*e*) None of the above.

86. Database amangement system are intended to :
(*a*) eliminate data redunancy
(*b*) establish relationships among records in different files
(*c*) manage files access
(*d*) maintain data integrity
(*e*) All of the above

87. Administrate supervision of database activities is the responsibility of the
(*a*) Database administrator
(*b*) DP Manager
(*c*) DB Manager
(*d*) DP administration
(*e*) None of the above

88. Data integrity control :
(*a*) is used to set upper and lower limits on numeric data
(*b*) requires the use of passwords to prohibit unauthorized access to the file
(*c*) has the data dictionary keep the data and time of last access last backup
(*d*) All of the above
(*e*) None of the above

89. The data dictionary tells the DBMS :
(*a*) what files are in the database
(*b*) what attribute are possessed by the data
(*c*) what these files contain
(*d*) All of the above
(*e*) None of the above

90. A data dictionary is a special file that contains :
(*a*) the names of all fields in all file
(*b*) the data types of all fields in all files
(*c*) the widths of all fields in all files
(*d*) All of the above
(*e*) None of the above

91. The physical location of record is determined by a mathematical formula that transforms a file key into a record location in :
(*a*) a tree file (*b*) an indexed file
(*c*) a hashed file
(*d*) None of the above

92. The logical data structure with a one-to-many relationship is
(*a*) network (*b*) tree
(*c*) chain (*d*) relational
(*e*) None of the above

93. Which of the following is not a logical database structure ?
(*a*) Tree (*b*) Relational
(*c*) Network (*d*) Chain
(*e*) None of the above

94. Which of the following is true of a network structure ?
(*a*) It is a physical representation of the data
(*b*) It allows a many-to-many relationship
(*c*) It is conceptually simple
(*d*) It will be the dominate database of the future
(*e*) None of the above

95. Which of the following is not an advantgaes of the database approach?
(*a*) Elimination of the data redundancy
(*b*) Ability to associate related data
(*c*) Increased security
(*d*) Program/data independence
(*e*) All of the above are advantages

96. In SQL, hih command(s) is (are) used to re-compile a stored function?
(*a*) SET FUNCTION
(*b*) SET STORED FUNCTION
(*c*) ALTER FUNCTION
(*d*) All of the above
(*e*) None of the above

97. In SQL, which commands is (are) used to redefine an index's future stirage allocation ?
(*a*) ALTER INDEX
(*b*) REDEFINE INDEX
(*c*) MODIFY INDEX
(*d*) DO INDEX
(*e*) None of the above.

98. Which of the following command(s) is (are) used to recompile a stored procedure in SQL?
(*a*) Compile Procedure
(*b*) Alter Procedure
(*c*) Modify Procedure
(*d*) All of the above
(*e*) None of the above

99. In SQL, which command is used to add a column/integrity constraint to table

(a) ADD COLUMN
(b) INSERT COLUMN
(c) MODIFY TABLE
(d) ALTER TABLE
(e) None of the above

100. Which command(s) is (are) used to redefine a column of the table in SQL?
(a) Alter Table
(b) Define Table
(c) Modify Table
(d) All of the above
(e) None of the above

101. Which command(s) is (are) used to enables/is able/drop an intedrity constraint in SQL?
(a) Define Table (b) Modify Table
(c) Alter Table (d) All of the above
(e) None of the above

102. In SQL, which command (s) is (are) used to enable/disable all triggers on a table?
(a) Alter Triggers (b) Alter Table
(c) Modify Triggers in Table
(d) None of the above

103. In SQL, which command(s) is (are) used to change a table's storage characteristics ?
(a) Alter Table (b) Modify Table
(c) Change Table (d) All of the above.

104. The ..... command is used to alloacte an extent for the table in SQL.
(a) Modify Allocates
(b) Modify Table
(c) Alter Table
(d) Redefine Table
(e) None of the above

105. In SQL, THE ALTER TABLESPACE command is used :
(a) to add/rename data files
(b) to change storage characteristic
(c) to take a tablespace on-line/off-line
(d) to being/end a backup
(e) None of the above.

106. In SQL, which command(s) is (are) used to nable/disable a database trigger ?
(a) Alter Trigger (b) Alter Database
(c) Alter Table (d) Modify Trigger
(e) None of the above

107. In SQL, which command(s) is (are) used to enable/disable a database trigger?
(a) MODIFY USER(b) CHANGE USER
(c) ALTER USED (d) All of the above
(e) None of the above

108. In SQL, the .... command is used to re-compile a view
(a) COMPILE VIEW
(b) DEFINE VIEW
(c) CREATE VIEW
(d) ALTER VIEW

109. In SQl, the .... command (s) is (are) used to create an idnex for a table or cluster
(a) CREATE INDEX
(b) MODIFY INDEX
(c) SET INDEX TO FILE
(d) All of the above
(e) None of the above

110. In SQL, which aommand(s) is (are) used to issue multiple CREATE TABLE, CREATE VIEW, and GRANT statements in a single transaction ?
(a) CREATE PACKAGE
(b) CREATE SYNONYM
(c) CREATE CLUSTER
(d) All of the above
(e) None of the above.

111. In SQL, which command (s) is (are) used to create a synonym for a schema object ?
(a) CREATE SCHEMA
(b) CREATE SYNONYM
(c) CREATE SAME
(d) All of the abive
(e) None of the above

112. In SQL, the CREATE TABLE is used :
(a) to create, a table, defining its columns, integrity constrains, and storage allocation

(*b*) to add a column/integrity constraint to a table
(*c*) to redefine a column
(*d*) to change a table's storage characteristics
(*e*) None of the above.

113. In SQL, the CREATE TABLESPACE is used :
(*a*) to create a place in the database for storage of scheme objects,
(*b*) to crate a database trigger
(*c*) to add/rename data files, to change storage
(*d*) All of the above
(*e*) None of the above.

114. In SQL, which command is used to create a database user ?
(*a*) AD USER TO DATABASE
(*b*) MY USER
(*c*) CREATE USER
(*d*) All of the above
(*e*) None of the above

115. In SQL, the CREATE VIEW command is used :
(*a*) to re-compile view
(*b*) to define a view of one or more tables or views
(*c*) to re-compile a table
(*d*) to create a trigger
(*e*) No one of the above

116. In SQL, which command is used to remove a stored function from the database ?
(*a*) REMOVE FUNCTION
(*b*) DELETE FUNCTION
(*c*) DROP FUNCTION
(*d*) ERASE FUNCTION
(*e*) None of the above

117. Which command is used to remove an index from the database in SQL?
(*a*) DELETE INDEX
(*b*) DROP INDEX
(*c*) REMOVE INDEX
(*d*) ROLLBACK INDEX
(*e*) None of the above

118. Which command is used to remove a table from the database in SQL?
(*a*) DELETE TABLE
(*b*) DROP TABLE (*c*) ERASE TABLE
(*d*) UNMATCHED TABLE
(*e*) None of the above

119. In SQL, GRANT command is used to
(*a*) allow user to access database
(*b*) allow user to create database
(*c*) grant system privileges, roles and object privileges to uses and roles
(*d*) choose auditing for specific SQL commands
(*e*) None of the above

120. In SQL, which command(s) is (are) used to remove rows from a table?
(*a*) DELETE (*b*) REMOVE
(*c*) TRUNCATE (*d*) Both (*a*) and (*c*)
(*e*) None of the above

121. In SQL, which command is used to add new rows to a table ?
(*a*) ALTER TABLE (*b*) ADD ROW
(*c*) INSERT (*d*) APPEND
(*e*) None of the above.

122. In SQL, which command is used to changes data in a table ?
(*a*) UPDATED (*b*) INSERT
(*c*) BROWSE (*d*) APPEND
(*e*) None of the above

123. In SQL, which command is used to select data in rows and column from one or more tables ?
(*a*) CHOOSE (*b*) SELECT
(*c*) LIST (*d*) BROWSE
(*e*) None of the above

124. In SQL, which of the following is not a data Manipulation Language Commands ?
(*a*) DELETE (*b*) SELECT
(*c*) UPDATE (*d*) CREATE
(*e*) None of the above

125. In SQL, which of the following is not a data definition language commands ?
(*a*) RENAME (*b*) REVOKE
(*c*) GRANT (*d*) UPDATE

126. In SQL, which command is used to make permanent changes made by statements since the beginning of a transactions?
(*a*) ZIP
(*b*) PACK
(*c*) COMMIT
(*d*) SAVE
(*e*) None of the above

127. In SQL, which statement is used to select employee number (EMP_NO) and Departmeny (DEPT) from (EMPLE) table whose salary (SALARY) is 8000?
(*a*) SELECT EMPNO, DEPT FROM EMPL = 8000
(*b*) SELECT EMPNO, DEPT FROM EMPL WHERE SALARY = 8000
(*c*) SELECT EMPNO, DEPT, SALARY = 8000 FROM EMPL,
(*d*) SELECT EMPNO, DEPT, FROM TABLE EMPL WHEN SALARY – 800;

128. In SQL, which commands is used to select only one copy of each set of duplicate rows
(*a*) SELECT DISTINCT
(*b*) SELECT UNIQUE
(*c*) SELECT DIFFERENT
(*d*) All of the above
(*e*) None of the above

129. Which command is used to select distinct subject (SUB) from the table (BOOK) ?
(*a*) SELECT ALL FROM BOOK
(*b*) SELECT DISTINCT FROM BOOK
(*c*) SELECT SUB FROM BOOK
(*d*) All of the above
(*e*) None of the above

ANSWERS

1. (*b*)	**2.** (*c*)	**3.** (*b*)	**4.** (*c*)	**5.** (*d*)	**6.** (*c*)	**7.** (*b*)	**8.** (*a*)
9. (*b*)	**10.** (*d*)	**11.** (*a*)	**12.** (*c*)	**13.** (*a*)	**14.** (*c*)	**15.** (*b*)	**16.** (*b*)
17. (*c*)	**18.** (*d*)	**19.** (*a*)	**20.** (*d*)	**21.** (*c*)	**22.** (*a*)	**23.** (*d*)	**24.** (*b*)
25. (*b*)	**26.** (*b*)	**27.** (*a*)	**28.** (*b*)	**29.** (*b*)	**30.** (*b*)	**31.** (*b*)	**32.** (*b*)
33. (*b*)	**34.** (*b*)	**35.** (*d*)	**36.** (*b*)	**37.** (*d*)	**38.** (*b*)	**39.** (*d*)	**40.** (*b*)
41. (*d*)	**42.** (*c*)	**43.** (*c*)	**44.** (*d*)	**45.** (*b*)	**46.** (*a*)	**47.** (*d*)	**48.** (*a*)
49. (*c*)	**50.** (*d*)	**51.** (*c*)	**52.** (*a*)	**53.** (*d*)	**54.** (*b*)	**55.** (*b*)	**56.** (*b*)
57. (*a*)	**58.** (*c*)	**59.** (*b*)	**60.** (*a*)	**61.** (*c*)	**62.** (*a*)	**63.** (*c*)	**64.** (*b*)
65. (*c*)	**66.** (*c*)	**67.** (*a*)	**68.** (*c*)	**69.** (*c*)	**70.** (*d*)	**71.** (*c*)	**72.** (*b*)
73. (*c*)	**74.** (*b*)	**75.** (*c*)	**76.** (*a*)	**77.** (*d*)	**78.** (*c*)	**79.** (*d*)	**80.** (*c*)
81. (*d*)	**82.** (*d*)	**83.** (*e*)	**84.** (*a*)	**85.** (*e*)	**86.** (*a*)	**87.** (*a*)	**88.** (*a*)
89. (*a*)	**90.** (*d*)	**91.** (*a*)	**92.** (*b*)	**93.** (*d*)	**94.** (*d*)	**95.** (*e*)	**96.** (*c*)
97. (*a*)	**98.** (*b*)	**99.** (*b*)	**100.** (*a*)	**101.** (*c*)	**102.** (*b*)	**103.** (*a*)	**104.** (*c*)
105. (*d*)	**106.** (*a*)	**107.** (*c*)	**108.** (*d*)	**109.** (*a*)	**110.** (*b*)	**111.** (*b*)	**112.** (*a*)
133. (*a*)	**114.** (*a*)	**115.** (*b*)	**116.** (*c*)	**117.** (*b*)	**118.** (*b*)	**119.** (*c*)	**120.** (*d*)
121. (*c*)	**122.** (*a*)	**123.** (*b*)	**124.** (*d*)	**125.** (*d*)	**126.** (*c*)	**127.** (*b*)	**128.** (*a*)
129. (*b*)							

FILL IN THE BLANKS

1. Processed data is known as ....
2. The process of collecting all items of data together to produce meaningful information is known as ....

3. In a table, columns represent .... and rows represent .... of data.
4. The .... is a column used to uniquely identify a row of data in a table and/or establish a relationship with another table.
5. Composite key is also known as .... or ....
6. The data, which describe the tables, columns, indexes, constraints and other items that makeup the database is known as ....
7. In a database, entities are represented by .... and attributes are represented by ....
8. The number of attributes associated with the table or relation is known as ....
9. .... refers to the duplication of data in a database whereas ....refers to ensuring that the data in the database is accurate.
10. The ability to modify a schema definition in one level without affecting a schema definition in the next higher level is known as ....
11. .... is a program, which provides the interface between the physical database and the application programs and queries submitted to the database system.
12. .... form of normalization removes transitive dependency.
13. Data .... is a process where the data in a data warehouse can be analysed to reveal hidden patterns and trends in historical business activity.
14. In the .... database model, relationship between record from a treelike structure.
15. A .... is defined as a database containing descriptions and definitions. Concerning the structure, data elements, interrelationships and other characteristics of a database.
16. SQL based queries collect rows having sense value in the .... clause in one cluster.
17. The language used for data modification is known as ....
18. .... Clause is used to eliminate some groups from a SELECT query.
19. The technique of passing the result of inner query to butter query is known as ....
20. The post of a software program that a user sees and interacts with is known as ....
21. SELECT statement describe .... to retrieve, rather than .... to retrieve.
22. The three main SQL standards are ...., ... and ....
23. Group Function are also known as .... functions.
24. Instead of specifying centre column list, can also be used to select all columns.
25. Full form of PL/SQL is ....

ANSWERS

1. Information	**2.** Data Processing	**3.** Fields, Records
4. Key	**5.** Concatenated key, structured key	
6. Metadata	**7.** Tables, fields	**8.** Degree
9. Data redundancy, Data integrity	**10.** Data independence	
10. Data independence	**11.** Third	**12.** Disk manager
13. Data mining	**14.** Hierarchical	**15.** Data dictionary
16. GROUP BY	**17.** DML	**18.** HAVING
19. Subquerry	**20.** Fronted	**21.** What, How
22. SQL 89, SQL 92, SQL 99	**23.** Aggregate	**24.** Asterisk (*)
25. Procedural language/Structured Query Language.		

TRUE OR FALSE

1. Knowledge is an ordered collection of facts, rules, understanding and principles that can be written down.
2. A conceptual schema defines the whole database without reference to how the data is stored.
3. Data and information are indistinguishable terms.
4. E.F. codd developed the hierarchical DSMS Model in 1960s.
5. A relationship is an association between attributes.
6. A user view is a logical discrimination of a portion of the database.
7. Nominalization is the process of decomposing relations to produce smaller, well-saturated tables.
8. In a client/server database system, the database generally resides on a server.
9. A distributed database requires multiple database management system, running at each remote site.
10. The hierarchical data model is the most commonly used model in contemporary database applications.
11. A composite key is an attribute in one table that acts as a primary key in another table.
12. In a decartelized database, the computer are not interconnected by a network and users cannot share data.
13. A logical grouping of characters is a field.
14. In SQL, columns of different tables can have the some name.
15. ODBS is an example of back end.
16. SQL join creates a new table by joining two or more tables.
17. SELECT name item amount FROM customer; is an invalid query.
18. SQL is owned jointly by FBM and Oracle.
19. The maximum limit of a VAR CHAR data type is 256 characters.
20. In SQL, SELECT, if there are more than one table after the FROM keyword, if represents a join.
21. SQL is case insensitive language.
22. The AND operator displays a record if any conditions listed true. The OR operator displays a record if ALL of the conditions listed are true.
23. The SQL REMOVE command deletes data from a database.
24. SQL is a Procedural Language.
25. The traditional approach to information processing is file-oriented.
26. A DBMS is a set pf programs that serve as an interface between application programs and a database.
27. A database is a collection of programs.
28. Database management systems are designed with a view towards optimizing the use of physical storage and CPU processing time.
29. The Database administrator creates and maintains edit controls, regarding changes and additions to the database.

30. The purpose of a DBMS is to provide a mechanism for organising, structuring and storing data.
31. Forms can only be used to enter new information.
32. Look-up, Insert, Delete, and Modify are primitive operations common to all record management systems.
33. List can be kept in order by indexing.
34. All record management system used the same cycle as a model of the information flow.
35. Hashed files are an example of a physical structure.
36. The hierarchy in a hierarchical schema cannot be changed easily once the database is set up.
37. A logical schema describe how data is stored on disk.
38. To use a logical schema. It is necessary to be a professional database administrator.
39. A network schema is more flexible than hierarchically schema.
40. The JOIN commands is used to merge two files.
41. The relations DBMS, the data structures are defined in logical, rather than physical terms.
42. Since a turnaround document is machine recardability is not considered a source of data.
43. Date element is another name for field.
44. Integrated databases are made possible by DBMS software.
45. The entire sequential file must be processes if only one record is to be updated.
46. A key data element is not needed for sequential processing.
47. One of the disadvantages of DBMS software is that application programs must be modified when the data design is changed.
48. All field in a record must be of the same type.
49. To use a subschema it is necessary to understand how data is stored on disk.
50. Each individual view of a database is called a subschema.
51. A new file created by adding a new entry into the data dictionary.
52. A query command is an instruction to the DBMS.
53. Queries are expressed in English.
54. A compound key is made up of a single piece of information.
55. The linking of data items within levels of a database structure is called changing.
56. The DBS implements security and disaster-recovery procedures.
57. The logical structure of a database is sometimes called a schema.
58. The Data/Definition Language (DDL) is used to define the physical characteristics of each record; the field within the record, and each field's name, data type, and length.
59. The DDL provides the techniques for processing the database, such as retrieval, sorting, display, and deletion of data or records.
60. The primary Function of the DDL is to provide the technique for data manipulation such as deletion, replacement, retrieval, sorting, or inserting of data.
61. The modify operatioin is used to add record to file.
62. In some DBMS programs, data compression is used to save disk space.

63. Queries can empty aggregate functions to computer an answer.
64. If data is stored redundancy, then data inconsistencic can occur.
65. Each row of a relation is called a domain.
66. A relational schema stores data in trees.
67. All DBMS files can be used by only one user at a time.
68. A locked file can be used by onle one user.
69. The function of a DML is to provide for the use of logical relationships among data items.
70. The function of DDL is to provide for logical and physical data independence.
71. The database administrator is, in effect, the coordinator between the database and the users.

ANSWERS

1. True	**2.** True	**3.** False	**4.** False	**5.** True	**6.** True
7. True	**8.** True	**9.** True	**10.** False	**11.** False	**12.** False
13. True	**14.** True	**15.** False	**16.** False	**17.** True	**18.** False
19. False	**20.** True	**21.** True	**22.** False	**23.** False	**24.** False
25. True	**26.** True	**27.** True	**28.** True	**29.** True	**30.** True
31. False	**32.** True	**33.** True	**34.** False	**35.** True	**36.** True
37. True	**38.** False	**39.** True	**40.** True	**41.** True	**42.** False
43. True	**44.** True	**45.** True	**46.** False	**47.** False	**48.** False
49. False	**50.** True	**51.** True	**52.** True	**53.** False	**54.** False
55. True	**56.** True	**57.** True	**58.** True	**59.** False	**60.** False
61. False	**62.** True	**63.** True	**64.** True	**65.** False	**66.** False
67. False	**68.** True	**69.** True	**70.** True	**71.** True	

❑❑❑

TUTORIAL EXERCISES

TUTORIAL EXERCISE No. 1

OBJECTIVE: Give an ER model for a University information system like NSS:

1. Maintain contact information for staff and students.
2. Provide handbook information for courses (syllabus, pre-reqs.UOC.)
3. Provide handbook information for programmes (requirement,schedule....)
4. Allow students to enroll in courses in classes (quotas, wait lists,...)
5. Allow school to book rooms for classes (whole semester, one-half,...)
6. Provide transcripts, progress check academic standing information.
7. Maintain marks for individual assignments as well as whole course (*i.e.*, SMS)

OBJECTIVE: Give an ER model for the following scenario:

1. Patients are identified by an SSN, and their name, addresses and ages must be recorded.
2. Doctors are identified by an SSN. For each doctor, the name, especially and years of experience must be recorded.
3. Each pharmacy by has a name, address and phone no. Pharmacy must have a manager.
4. A pharmacist is identified by an SSN. He/she can only work for one paharmacy. For each pharmacist, the name, qualification must be recorded.
5. For each drug, the trade name and formula must be recorded.
6. Every patient has a primary physician. Every doctor has at least one patient.
7. Each pharmacy sells several drug, and a price for each. A drug could be sold at several pharmacies, and the price could vary between pharmacies.
8. Doctors prescribe drugs for patients. A doctor could prescribe one or more drugs for several patients, and a patient could obtain prescription from several doctors. Each prescription has a date and quantity associated with it.

OBJECTIVE: Produce an ER design and then relational design for:

1. For each customer, we record name, address, customer id.
2. For each employee, we record name, address, payroll no.
3. For each branch, we record its name and address.
4. The bank provides each customer with personal banker.
5. Each employee works under a manager in branch.
6. There are saving accounts and cheque accounts.
7. Saving have interest; cheques have overdraft limit.
8. Accounts are held centrally.
9. Customers can have any number of accounts.
10. Accounts can be held jointly.
11. Record last, access data for each account holder.
12. Each customer can take out one loan individually.
13. Loans are associated with a particular branch.
14. Loan payments are associated with a give loan.

❑❑❑

TUTORIAL EXERCISE No. 2

OBJECTIVE: Consider the following variant of the employee-department-project schema:

Emp (eid,bdate,name,did) (did=worksfor)

Dept(name,phone,eid) (eid=manager)

Project(pid,name,budget)

Partic(eid,pid,time)

Depend(eid,name,bdate,reln)

Give relational algebra expressions to answer the following:

1. Who has the most dependent?
2. What is the phone no. of the sales dept.?
3. Which projects have a budget of more than $1,000,000?
4. Which employees were born after 1980?
5. Who is the manager of the research dept.?
6. Who is working on the chemical weapons project?
7. Does anyone work on all of the projects?
8. Who has a dependent called Timmy?

OBJECTIVE: Consider the following relational database and answer the following queries in Relational Algebra.

Employee (ename, street, city)

Works (ename, cname, salary)

Company (cname, city)

Manages (ename, mname)

1. Find the names of all employees who work for first Bank Corporation.
2. Find the names and cities of residence of all employees who works for first Bank Corporation.
3. Find the names, street address and cities of residence of all employees who works for first Bank Corporation and earn more than 10,000 per annum.
4. Find the name of all employees in this database who live in the same city as the company for which they work.
5. Find the names of all employees who live in the same city and on the same street as do their managers.
6. Find the name of all employees in this database who do not work for first Bank Corporation.
7. Find the name of all employees in this database who earn more than every employee of small bank corporation.
8. Assume the companies may be located in several cities. Find all companies located in every city in which small bank corporation is located.

❑❑❑

TUTORIAL EXERCISE No. 3

OBJECTIVE: Given the relational schemes:

Enroll(S#, C#, Section), Advise(Prof, S#)

Teach (Prof, C#, section), student(S#, sname)

Give queries expressed in relational algebra, tuple calculus

1. List all students taking courses with smith or Jones.
2. List all students taking at least one course that there advisor teaches.
3. List all students who is advised by "Dr. Jone".

OBJECTIVE: Consider the following relational data base and answer the following queries in relational tuple calculus and relational domain calculus

Emp(ename, street, city)

Works (ename, cname, salary)

Company (cname, city)

Manages (ename, mname)

1. Find the names of all employees who work for "FBC"
2. Find the names and residence of all employees who work for FBC.
3. Find the names, street and cities of residence of all employees who work for FBC and earn more than 10,000 per annum.
4. Find the names of all employees who live in the same city on the same street as do their managers.
5. Find the names of all employees who live in the same city as the company for which they work.
6. Find the name of all employees who do not work for FBC.
7. Assume the companies may be located in several cities. Find all companies located in every city in which "SBC" is located.

❑❑❑

TUTORIAL EXERCISE No. 4

OBJECTIVE: Consider the relational database

Emp (**pname**, street, city)

Works (**pnarne**, cname, salary).

Company (**cname**, city)

Manages (**pname**, mname)

Where the primary keys are underlined. Give an expression in relational algebra to express each of the following queries:

1. Find the names of all employees who work for "FBC".
2. Give all the managers in this database a 10% salary raise.
3. Find the company with the most employees.
4. Find those companies whose employees earn a higher salary, on average than the average salary at "FBC".
5. Give all employees of "FBC" a 10% salary raise.
6. Delete all tuple in the works relation for employees of small bank corporation.
7. Modify the database so that Jones now lives in new town.

OBJECTIVE: Consider the following relations for a database that keeps track of student's enrollment in courses and the book adopted for each course.

Student (**ssn**. name, major, bdate)

Course(**c#, c**name, dept)

Enroll (**ssn. c#,** quarter, grade)

Book Adoption(c#, Quarter, book isbn)

Text (book isbn, book, title, publisher, author)

The underlined, attributes from the primary key of a relational schema specify the following queries in relational algebra.

1. List the no of courses taken by all students named "John smith" in Winter 2002 (*i.e.*, quarter = w02).
2. Produce a list of text books (include c# book ISBN, title) for courses offered by the CS dept. that have used more than two books.
3. List any dept. that has all its adopted books published by 'B.C.Publicating'.

TUTORIAL EXERCISE No. 5

OBJECTIVE: Consider the relational database:

Employee (eid,emp-name,emp-street,emp-city)

Works (eid,cid,salary)

Located in (c code,cname,ccity).

Write the following queries in SQL.

1. Find the names of the employees who work in company "sanamed".
2. Find the names of all employees who work for a company located in "Lucknow".
3. Find the name of the employees who do not work for company "sanamed".
4. Find the name of the employees who work for the company located at city in which they live.
5. Find the address (street, city) of an employee who gets the salary more than Rs. 10000 and is working in the company "sanomed".

OBJECTIVE: Consider the following scheme:

Supplier (s_id, s_name, s_address)

Parts (p_id, p_name, color)

Catalog (s_id, p_jd, cost).

Write the following queries in relational algebra and SQL.

1. Find the name of the suppliers who supply yellow parts.
2. Find the name of suppliers who supply both blue and green parts.
3. Find the name of suppliers who supply all parts.

❐❐❐

TUTORIAL EXERCISE No. 6

1. Construct an ER diagram for a hospital with a set of patient and a set of medical doctors. Associate with each patient log of the various tests and examination conducted.

OBJECTIVE: Consider the following set of requirements for a bank data base: "A large bank has several branches and different places. Each branch maintains the account details of the customers. The customer may open joint as well as single accounts.

The bank also provides the bank loans to the customers for different purposes. Bank keeps records of each transaction by the customer to his account. All of the branches have employee and some employees are managers.

2. Draw an ER diagram that captures this information.
3. Transform this ER diagram to the relational database scheme.
4. Write SQL-DDL statement to implement bank data base schema
5. Construct an ER diagram for a car insurance company whose customers own one or more cars each. Each car has associated with it zero to any no. of recorded accidents.
6. Give an ER diagram for showing fatherhood, motherhood and his spouse relationship among man woman and construct the appropriate table for this ER diagram.

❑❑❑

TUTORIAL EXERCISE No. 7

1. What is functional dependency? List all functional dependencies satisfied by relation show below.

A	B	C
a1	b1	c1
a1	b1	c2
a2	b1	c1
a2	b1	c3

2. Given a relation R(x,y,z,w) and the set of FDs is f{io→x,x→y,wo→ xy}. Find the closure set of functional dependency.
3. Given R(A,B,C,D,) with the functional dependencies f {ABH → CD, C → D}, the composition of R into Rl (A,B,C),R2(B,C,D) is loss less join decomposition or not. Justify your answer.
4. What do mean by dependency preserving? Consider the relation R (A, B, C, D) with functional dependency f{A → B, A → C, C → D} and its decomposition into R1(A, B, C and R2(C, D).Show that this decomposition is dependency preserving or not. It is a lossless join decomposition? Justify your answer.

OBJECTIVE: Consider the scheme R= (A,B,C,D), suppose following FDs hold:

E → A

CD → E

A → BC

B → D

State whether the following decomposition of R lossless join decomposition or not

5. {(A,B,C),(A,D,E)}
6. {(A,B,C),(C,D,E)}

❑❑❑

TUTORIAL EXERCISE No. 8

1. Given R(A,B,C,D,E,F) and the set of FDs on R given by F= {ABC → DE, AB → D, DE → ABCF,E → C}, in what normal form is R ? If it is not in 3NF, decompose R and find a set of 3NF projections of R. Is this set lossless and dependency preserving?
2. Decompose the relation R into 4NF.

   R(A,B,C,D,E,F,G)

   M { A → B, B → G, B → EF, CD → E}
3. Given a relation R(A,B,C,D,E) and a set F{AB → C, AB → D, D-H → A, BC → E }. Is this relation in BCNF? If not decompose the relation R into BCNF.
4. Consider the relation

   Car-sale (car#, date sold, salesman#, commission, discount amt salesman# commission.

   Is this relation is INF, 2NF or 3NF? Why or why not? How would you excessively normalize it completely?

OBJECTIVE: Consider the universal relation R(A,B,C,D,E,F,G,H,I,J) and the set of function dependencies F as given below:

F = {AB → C,A → DE, B → F F → GH,D → IJ}

5. Determine the key for R.
6. Decompose R into second Normal form.

❑❑❑

TUTORIAL EXERCISE No. 9

OBJECTIVE: Consider the following two transaction:

TI: Read(A)
Read (B)
If A = 0 than B = B + I
Write (B)
T2 : Read (B)
Read (A)
If B = 0 than A = A + I
Write (A)

1. Show a concurrent execution of T1 and T2 that produces a non-serializable schedule.
2. Is there a concurrent execution of T1 and T2 that produces a non-serializable schedule?
3. Show that every serial execution involving these two transactions preserves the consistency of database.
4. Show whether the following schedules are conflict equivalent or not. Justify your statement:

Schedule 1		Schedule 2	
TI	T2	TI	T2
Read (A)			
Writer (A)			Read (A)
	Read (A)	Read (A)	
	Write (A)	Write (A)	Write (A)

5. State the condition when two schedules are considered as view equivalent. State, whether the following schedule is view serializable. Justify your answer.

TI	T2	T3
Read (A)		
	Write (A)	
Write (A)		
		Write (A)

❒❒❒

TUTORIAL EXERCISE No. 10

OBJECTIVE: Which of the following schedule is conflict serializable? For each serializable schedule, determine the equivalent serial, schedules.

1. r1(x); r3(x); w1 (x);r2 (x);w3(x);
2. rl(x);r3(x);w3(x);wl(x);r2(x);
3. r3(x);r2(x);w3(x);rl(x);wl(x);
4. Consider the three transactions Tl, T2 and T3 and the schedules S1 and S2 given below. Draw the serializability (precedence) graph for S1 and S2.

   T1 :r1(x);rl(z);w1(x);

   T2 : r2(z);r2(y);w2(z);w2(y);

   T3 : r3(x);r3(y);w3(y)

   S1 : rl(x);r2 (z);rl(z);r3(x);r3(y);w3(y);r3(y);w2(z);w2(y)

   S2 : rl(x); r2 (z);r3(x);rl(z);r2(y);r3(y);wl(x);w2(z);w3(y);w2(y);

Tl	T2	T3
Read (y)		Write (y)
Write (y)	Read (z)	Read (x)
Read (x)	Write(y)	Write (z)

5. Show that the two phase locking protocol ensure conflict serializability, and that transaction can be serialized according to their lock points.

❑❑❑

PREVIOUS YEAR QUESTION PAPER

End Term Examination
Fourth Semester (B.Tech.), May-June, 2014
CONCEPT OF DATABASE SYSTEM

Time: 3 hours **Maximum Marks: 75**

Note : **Answer any five Questions including Q. no. 1 which is compulsory.**

1. Attempt **any five** questions:

(*a*) Explain the purpose of database system.

(*b*) Explain Timestamp based protocols.

(*c*) Differentiate between DBMS and file systems.

(*d*) Explain the basic structure of SQL.

(*e*) What are anomalies in database design?

(*f*) Explain the advantage of object oriented databases.

(*g*) What is shadow paging? Write any two disadvantages of it.

2. (*a*) What is data dependence? Explain the difference between logical and physical data dependence with example. (7.5)

(*b*) Discuss the basic concepts and architecture of database systems. (7.5)

3. (*a*) Discuss the various aggregate functions in SQL with an example for each. (7.5)

(*b*) Construct E-R diagram of the bank. It provides different kinds of bank accounts and loans. Also it operates number of branches. (7.5)

4. (*a*) Define the terms: Ternary relationship, Weak entity set, Identity owner, Aggregation.

(*b*) What is a Relation? Differentiate between a Relation schema and a Relation instance. Define the terms cardinality of a relation. (7.5)

5. (*a*) What is normalization? Why we need normalization of DBMS? (7.5)

(*b*) Explain about 2NF, 3NF and BCNF with an example for each. (7.5)

6. (*a*) What is deadlock? When it is occurs and how to avoid it? (7.5)

(*b*) Explain measures of query cost and selection operation in query processing. (7.5)

7. (*a*) What is meant by file organization and indexing? Discuss the Hash-based indexing. (7.5)

(*b*) Describe the concepts Serializability and Recoverability. (7.5)

8. Write short notes on the following :

(a) Data definition in SQL.

(b) Oracle 8 Architecture

(c) Distributed DBMS.

End Term Examination
Fourth Semester (B.Tech.), May-June, 2016
CONCEPT OF DATABASE SYSTEM

Time: 3 hours ***Maximum Marks: 75***

***Note :* Attempt any five questions including Q.No. 1 which is compulsory.**

1. Define the following (Give example where necessary):
   (*a*) Multivalued dependency (*b*) Dependency preservation
   (*c*) Functional dependency (*d*) Referential integrity
   (*e*) Internal schema (*f*) Candidate key
   (*g*) Triggers (*h*) DBA
   (*i*) Weak entity (*j*) Foreign key

2. (*a*) Differentiate between the term Generalization and Specialization with example. (6.5)
   (*b*) Differentiate between Fragmentation, Replication and Transparency. (6)

3. (*a*) Define closure of FD set. For the following relation :
   R (*A*, *B*, *C*, *D*)
   with FDs as follows :
   (i) AB → C
   C → A
   BC → D
   ACD → D
   D → EG
   BE → C
   CG → BD
   CE → AC
   Find the closure of (B, D) and (C, A). (6.5)
   (*b*) What is canonical cover? How it is computed? (6)

4. (*a*) Explain the "ACID" properties in brief. (6)
   (*b*) What do you mean by serializability. Discuss the conflict serializability and view serializability with examples. (6.5)

5. (*a*) Draw and explain a neat diagram of three level architecture of database system. (6)
   (*b*) Construct an ER diagram of 'Hospital management system'. There would be a set of patients and set of doctors. For each patient, there would be a log of various tests and examinations conducted. Make assumptions if necessary and clearly state them. (6.5)

6. (*a*) How the B+ tree index files are maintained? Explain. (6)
   (*b*) Explain the method of "query processing" in brief. (6.5)

7. (*a*) Discuss clearly inner join, outer join and theta join with example. (6)
   (*b*) For the relation R = (A, B, C) and S = (D, E, F) and relation r(R) and s(s), give SQL statement for following expressions and explain. (6.5)
   (i) $\pi_A(r)$ (ii) $\sigma_B = 17(r)$
   (iii) $r X s$ (iv) $\pi_{A,\ F}(\sigma_{C\,=\,D}(rXs))$

8. Attempt any two : (2×6.25=12.5)
   (a) How the deadlock is detected in transactions? Explain its recovery process also.
   (b) Normalize the following relation to as much as possible citing the reasons and anamoloies. R(emp-no, name, street, city, compangname, company-city, manager-name, age, salary, marital-status, spouse-name)
   (c) State the multi-version time stamp based protocol. Suggests a scheme to avoid the phantom phenomenon.

❑❑❑

INDEX

G

H

I

J

K

L

M

N

O

P

Q

R

S

T

U

V

W

❑❑❑

Win Prizes

ATTENTION READERS

Win Prizes

We request for your frank assessment regarding various aspects of the book as given below :

CONCEPTS OF DATABASE SYSTEM

(FOR GGSIPU)

By: P. K. YADAV

Please Fill up the Sheet in Neat and Clean Writing

(i) What topics of your syllabus, important for your examination point of view, are not covered in this book ?

....

....

....

....

(ii) What are the chapters wherein the treatment of the subject matter is not systematic or organised or updated ?

....

....

....

....

(iii) Have you come across misprints/mistakes/factual inaccuracies in the book? Please specify the chapters and the page numbers.

....

....

....

....

(iv) What other books on the same subject have you found/heard better than the present book? Please specify in terms of price and quality

....

....

....

....

(v) Further suggestions and comments for the improvement of this book.

....

....

....

....

(PS: If need be please attach a separate sheet to write your views)

OTHER DETAILS

(i) Who recommended you this book ? (Please tick (✓) in the box near the option relevant to you.

☐ Teacher ☐ Friends ☐ Bookseller

....

....

....

....

(ii) Name of the recommending teacher, his designation and address :

....

....

....

....

(iii) Name and address of the bookseller you purchased the book from :

....

....

....

....

(iv) Name and address of your institution (Please mention the University or Board as the case may be)

....

....

....

....

(v) Your name and complete postal address :

....

....

....

....

(vi) Course you are enrolled in :

....

....

....

....

The best assessment will be awarded each month: The award will be in the form of our publications as decided by the Editorial Board.

Please mail the filled up coupon at your earliest to :

S.K. KATARIA & SONS®

4885/109, Prakash Mahal, Dr. Subhash Bhargav Lane,

Opposite Delhi Medical Association, Daryaganj, New Delhi–110002 (INDIA)

Phone: +91-11-23243489, 43551243; *Mobile:* +91-9871775858

e-mail: katariabook@gmail.com, katariabooks@yahoo.com

Website: www.skkatariaandsons.com